The Potential of Oil Palm Solid Waste in Industrial Processes to Support Sustainable Industrial Growth

Theory and Case Studies

nova
science publishers
www.novapublishers.com
New York

Environmental Science, Engineering and Technology

More information about this series can be found at

Chemical Engineering Methods and Technology

More information about this series can be found at

Renita Manurung
Hamidah Harahap
Rosdanelli Hasibuan
Halimatuddahliana Nasution
Sri Zahrani Dwi Mauliyah Parinduri
Ayuni Yustira
Viqry Pramananda

The Potential of Oil Palm Solid Waste in Industrial Processes to Support Sustainable Industrial Growth

Theory and Case Studies

nova
science publishers
www.novapublishers.com
New York

DOI: https://doi.org/10.52305/IVTQ7111

NOTICE TO THE READER

Library of Congress Cataloging-in-Publication Data

ISBN: 979-8-90134-327-2 (Hardcover)
ISBN: 979-8-90134-360-9 (eBook)

Published by Nova Science Publishers, Inc. † New York

This book is dedicated to the academic community of Universitas Sumatera Utara (USU), where the pursuit of knowledge and innovation is fostered through a spirit of collaboration and excellence. We extend my deepest appreciation to my colleagues, students, and mentors at USU for their continuous inspiration, critical insights, and unwavering support throughout the development of this work. This book reflects our shared commitment to advancing sustainable industrial practices and contributing to the global knowledge base. May this work serve as a valuable resource for future generations of scholars and practitioners, helping to pave the way for a more sustainable and innovative future.

Contents

Description

The Potential of Oil Palm Solid Waste in Industrial Processes to Support Sustainable Industrial Growth: Theory and Case Studies provides a thorough and in-depth exploration of the valorization of oil palm solid residues within the broader context of sustainable industrial development. As one of the most significant pillars of the bio-based economy in tropical regions, the palm oil industry inevitably generates substantial quantities of solid biomass at both plantation and processing stages. This monograph systematically examines the scientific, technological, and industrial pathways for converting these residues - including empty fruit bunches (EFB), mesocarp fibers, shells, fronds, trunks, and leaves - into high-value materials that reinforce a circular, resource-efficient industrial ecosystem.

A diverse range of utilization strategies is discussed, beginning with the biochemical conversion of EFB and mesocarp fibers into bioethanol through fermentation, and extending to the transformation of shells and other lignocellulosic fractions into advanced adsorbents and catalysts applicable to wastewater treatment and biodiesel synthesis. The book also highlights the increasing relevance of thermochemical and physicochemical technologies in producing bio-oil, syngas, biochar, and silica-based catalysts. These processes not only elevate the economic value of palm biomass, but also create new opportunities for industrial innovation, product diversification, and enhanced sustainability within the palm oil sector.

Beyond technological considerations, the monograph underscores the broader economic, environmental, and policy implications of biomass utilization. Transforming waste into value-added materials can significantly reduce waste management burdens, mitigate carbon emissions, and reinforce circular economy initiatives in major palm-oil-producing nations such as Indonesia.

Overall, this monograph positions palm-based biomass as a strategic and underutilized resource for achieving sustainable industrial growth. Through the integration of theoretical perspectives and practical case studies, it bridges scientific knowledge with industrial applications, contributing to the development of greener, more efficient, and more resilient production systems across the palm oil industry.

Preface

Sustainable industrial growth is a critical goal for the global economy, especially in the face of mounting environmental challenges. As industries strive to reduce their environmental footprint, the potential of industrial waste - specifically from agriculture - emerges as a key area for innovation. Among the most significant sources of waste in the agricultural sector is the oil palm industry, which produces large quantities of solid waste, such as empty fruit bunches, palm kernel shells, and palm oil mill effluent. Despite the substantial volume of waste generated, these materials remain underutilized.

The purpose of this book is to explore the multifaceted potential of oil palm solid waste in industrial processes. By integrating both theoretical frameworks and real-world case studies, this work aims to demonstrate how these waste products can be transformed into valuable resources, contributing to the dual objectives of sustainable industrial growth and environmental stewardship. The chapters within this book present a range of innovative applications, from bioenergy production to the development of bio-based materials, underscoring the importance of waste valorization in achieving sustainability goals.

This book is intended to serve as a comprehensive resource for academics, industry professionals, and policymakers who are interested in exploring the potential of oil palm waste and its role in driving sustainable industrial practices. The case studies presented offer practical insights into the technologies, processes, and strategies that can lead to more sustainable and economically beneficial outcomes. It is my hope that this work will inspire further research, collaboration, and action in the pursuit of a sustainable future.

Acknowledgments

The creation of this monograph would not have been possible without the support and collaboration of many individuals and institutions. First and foremost, we would like to extend our sincere gratitude to our academic mentors and colleagues, particularly Board of Professors Universitas Sumatera Utara (Dewan Guru Besar Universitas Sumatera Utara), whose guidance, encouragement, and expertise have been invaluable throughout the development of this book. We would like to express our deepest appreciation to Professor Mohammad Basyuni, S.Hut., M.Si., Ph.D., whose continuous support and invaluable direction, especially in administrative matters, played a crucial role in helping us complete this book. His mentorship from the very beginning of this project has been essential in ensuring its successful completion. Lastly, we would like to express our deepest gratitude to our families and friends for their constant support and encouragement. Your patience, understanding, and belief in the value of this work have provided us with the motivation to see this project through to completion.

Chapter 1

Oil Palm Solid Waste Potential in Indonesia

Abstract

Oil Palm is one of the most significant agricultural crops globally, especially in Indonesia, which leads in its production. The oil palm, *Elaeis guineensis*, not only provides essential oils for various industries but also generates substantial solid waste by-products during its processing. These by-products, including empty fruit bunches (EFB), palm kernel shells (PKS), mesocarp fiber, and palm fronds, have long been underutilized, despite their considerable potential for creating value-added products. With an emphasis on EFB and other residues, this chapter explores the extent of palm oil solid waste in Indonesia and emphasizes its potential for sustainable usage. Industries can improve the economic sustainability of the palm oil industry and drastically lessen their negative effects on the environment by processing these by-products. Oil palm trash can be transformed into biodegradable products, catalysts, adsorbents, renewable energy, and other novel materials using cutting-edge technologies. With an emphasis on the economic and environmental advantages of this waste-to-value strategy, the chapter provides a thorough review of the potential and problems currently facing the management and use of oil palm solid waste.

1. Introduction

Elaeis guineensis Jacq, the most important variation in the Elaeis genus and a member of the *Palmae* family, is the scientific name for the oil palm, a monocot plant (Janaun et al., 2010). The second kind is the American oil palm, which is native to South and Central America and is scientifically known as *Elaeis oleifera* (H.B.K) Cortes. Compared to the African variety, *Elaeis oleifera* has a lower oil production per bunch and a higher concentration of unsaturated fatty acids. In order to create interspecific hybrids that are of greater quality, these two types have been combined. An oil palm tree is photographed in Figure 1.

Figure 1. Oil palm tree.

The oil palm is a monoecious, erect plant that separates male and female inflorescences. It is cross-pollinated, and the primary pollinator is the beetle *Elaeidobius kamerunicus* Faust. Previously, wind pollination was used but yielded poor results. Assisted pollination was introduced to address the low pollination rates in oil palms. This practice became standard management for plantations. But once it was discovered that insects pollinate oil palms and a new variety, *E. kamerunicus*, was introduced from Cameroon, West Africa, in 1982, it was eventually abandoned. 24 to 30 months after planting, the majority of oil palm trees are harvested. Eight to fifteen fresh fruit bunches (FFB) with an average weight of 15 to 25 kg are produced annually by each tree.

The age of the tree and the kind of palm seed determine how much fruit may be harvested from oil palm trees. There are roughly 1000–1300 little fruits in each bunch. A fibrous mesocarp, or meat layer, and an endocarp, or skin, which contains the kernel, make up each little fruit. Presently, an average of 8.6 tons of crude palm oil (CPO) and palm kernel oil (PKO) are produced from an estimated 30 tons of FFB per hectare. Five to six tons of CPO are thought to be produced from thirty tons of FFB by the best oil palm trees. In a recent development, MPOC published results showing a 2013 CPO production of 5 tons and 20 tons of FFB per hectare. Fruit pigmentation and other traits set apart the different cultivars or breeding types of E. guineensis. Depending on the thickness of the endocarp and the amount of mesocarp, the most popular cultivars are Dura, Tenera, and Pisifera. The Dura palm has a medium mesocarp content (35–55% of the fruit weight) and an endocarp thickness of 2–8 mm, whereas the Tenera cultivar has a mesocarp content of 60–95% with an endocarp thickness of 0.5–3 mm. About 95% of the Pisifera variety is made up of mesocarp rather than endocarp.

In general, key aspects of oil palm tree maintenance involve increasing FFB yield and oil content, enhancing palm oil (CPO) quality, slowing tree height growth, and reducing disease infection. Nowadays, growing and producing highly marketed oil palm trees to satisfy market and industrial demands has been the main emphasis of the palm oil business (Bicalho et al., 2016). These are a few instances of enhancements that satisfy industry standards to support environmentally friendly and sustainable oil palm production. For example:

- Shorter oil palm varieties were introduced to decrease tree height growth and greatly increase the economic planting cycle.
- Seeds with larger amounts of unsaturated fatty acids are produced by cross-pollinating trees, such as the Nigerian Duras and Elaeis guineensis with Elaeis oleifera hybrids and maintaining hybrid oil palm varieties for high unsaturated palm oil trees (high iodine value).
- Creating seeds from certain Dura and Pisifera trees in Nigeria, breeding for high carotenoid content, and hybridizing with Elaeis oleifera.
- Using high-yielding Nigerian Dura oil palms with a high seed content to breed for a high lauric oil content.

1.1. Palm Oil Processing

A crucial phase in the palm oil industry's value chain is palm oil processing. Raw materials are transformed into primary products like crude palm oil (CPO) and other important by-products through a sequence of processes. Crude palm oil (CPO) and palm kernel oil (PKO), the primary products of palm oil processing, are used extensively in the food sector, cosmetics, and biodiesel manufacturing. The process of transesterification, which produces biodiesel, is one prominent use. According to Janaun et al., (2010), transesterification is the process of triglycerides reacting chemically with alcohol, usually methanol or ethanol, in the presence of a catalyst to create methyl esters, or biodiesel, and glycerol as a byproduct. Three long-chain fatty acids joined a glycerol molecule to make up triglycerides. Glycerol and biodiesel are produced when the alcohol or catalyst breaks the triglyceride bonds during the process. According to Yang et al., (2015), glycerol accounts for approximately 10% (w/w) of the biodiesel production output. In practical terms, for every gallon of biodiesel produced, approximately 1.05 pounds of glycerol are generated. For example, an annual production of 30 million gallons of biodiesel yields around 11,500 tons of 99.9% of pure glycerin.

As shown in Figure 2, oil palm demonstrates great potential as a renewable energy source, with various applications for producing biodiesel and other clean energy. This figure illustrates the processing flow of fresh oil palm fruit bunches through the palm oil mill. The process results in various products, such as crude palm oil, palm kernel oil, and liquid waste, which can be further processed into biodiesel. Additionally, the biomass produced, such as palm kernel cake, fronds, shells, fibers, and empty fruit bunches, can also be utilized for other products such as biofuel, briquettes, adsorbents, and bioethanol.

Otherwise, one key application of palm oil in renewable energy is biodiesel production. Figure 3 shows a simplified schematic of the basic transesterification process, which converts various oils - including palm oil - into biodiesel. This process reacts to the oils with an alcohol (typically methanol or ethanol) in the presence of a catalyst. Transesterification yields biodiesel (as methyl or ethyl esters) and glycerol as a byproduct. The resulting biodiesel serves as a renewable, environmentally friendly alternative to fossil fuels, while glycerol can be further refined for other industrial uses.

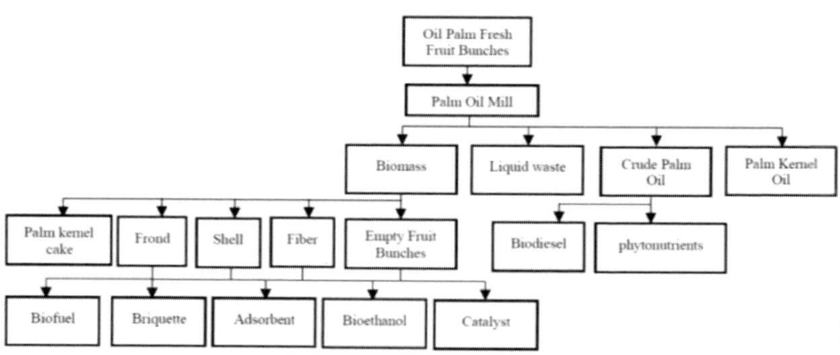

Figure 2. The potential of oil palm as a renewable energy source.

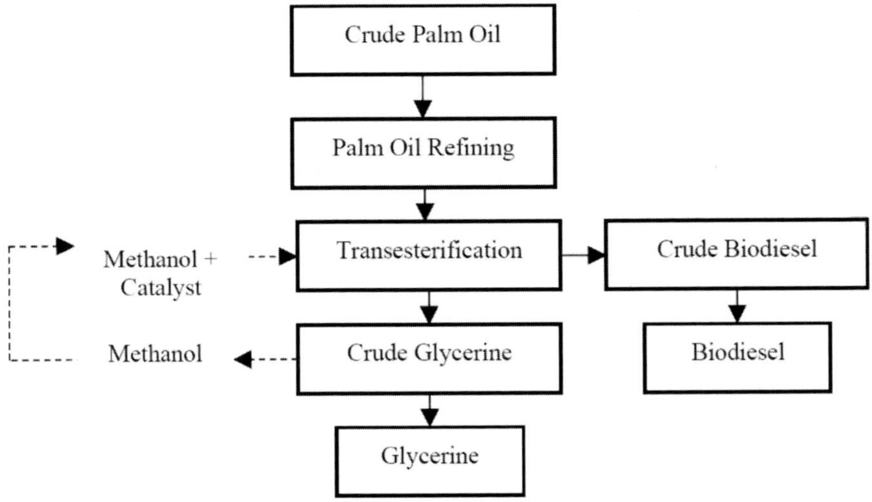

Figure 3. A simplified schematic diagram of the basic transesterification process used to convert various types of oils, including palm oil, into biodiesel.

1.2. The Potential of Oil Palm Solid Waste

Oil palm (*Elaeis guineensis*) is a crop that not only provides the primary raw material for palm oil production but also generates various waste and by-products that have often been regarded as mere waste. Palm oil production in Indonesia is highly significant, making the country one of the leading producers of palm oil globally. As shown in Figure 4, Indonesia contributes around 59% of the global palm oil production, with Malaysia following at 25%. The high level of production directly correlates with the substantial amount of waste generated. Currently, with over 14 million hectares of plantations, primarily in Sumatra and Kalimantan, Indonesia stands as the world's largest producer and exporter of palm oil, which poses a significant challenge in terms of disposal and management (Zainal et al., 2025). The waste generated includes empty fruit bunches, shell, fronds, and palm leaves. Without proper management, these wastes can become a serious environmental burden, leading to pollution, soil and water contamination, and greenhouse gas emissions that harm ecosystems. Recent studies (Taib et al., 2025) review the mechanical-chemical properties for high-value materials, while Budianta and Gozan (2022) proposes zero-waste through composting and energy. The government's 2025 policy aims to reduce POME/solid waste via PKS briquettes (1.2 MW electricity per mill in Lampung). This will enhance economic efficiency and reduce pollution.

As seen in Figure 5, oil palm biomass from plantations all around Indonesia has enormous potential for processing into high-value goods like fuels, carbon black feedstock, catalysts, and adsorbents. Considering that Indonesia produces more than half of the world's palm oil, this potential is significantly more than what is shown. Approximately 256 million tons of fresh fruit bunches (FFB) are produced year by palm oil plantations for use in palm oil mills. Additionally, 128.9 million tons of oil palm fronds (OPF) are produced through routine pruning, and 59.7 million tons of oil palm trunks (OPT) are harvested from mature plants (Aprianti et al., 2023). Severalby-products are also produced during the conversion of FFB into crude palm oil (CPO), such as 11–15% mesocarp fiber (PMF), 5.2–7% palm kernel shells (PKS), and 22–23% empty fruit bunches (EFB) (Aprianti et al., 2023). Sadly, most of this biomass is still unused, eventhough it may provide value and lessen its detrimental effects on the environment if treated appropriately.

In 2019, palm oil mills in Indonesia produced roughly 59 million tonnes of EFB, 17 million tonnes of PKS, and 33 million tonnes of mesocarp fiber. It is possible to efficiently use by-products like mesocarp fiber and palm shells to produce electricity, which may then be used to power the palm oil mills and employee housing. To encourage the growth of oil palm, OPT and OPF are usually left in plantation areas as extra fertilizer. Additionally, to avoid an excessive accumulation of waste at the mills, the EFB produced is returned to planting areas. Making use of this biomass benefits the palm oil sector financially and helps manage trash in a more environmentally friendly manner (Aprianti et al., 2023). Thus, it is crucial to create technology that can turn waste palm oil into more valuable and useful products, which will lessen its negative effects on the environment and increase the palm oil industry's financial gains.

In general, waste from palm oil mills consists of three types: liquid, solid, and gas waste. The solid waste from processing includes empty fruit bunches (EFB), shells, fiber, sludge, and cake (Waigaonkar et al., 2011). Palm oil production generates solid waste in the form of EFB (23%), palm shell waste (6.5%), palm sludge (4%), fiber (13%), and liquid waste (5%). Oil palm waste can provide significant benefits to society, such as organic fertilizer and activated charcoal production.

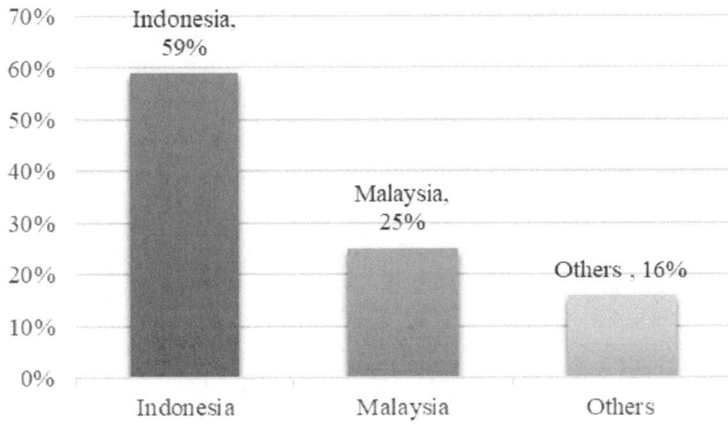

Figure 4. Top palm oil producers in thousand metric tons in 2021/2022 (Uchegbulam et al., 2022).

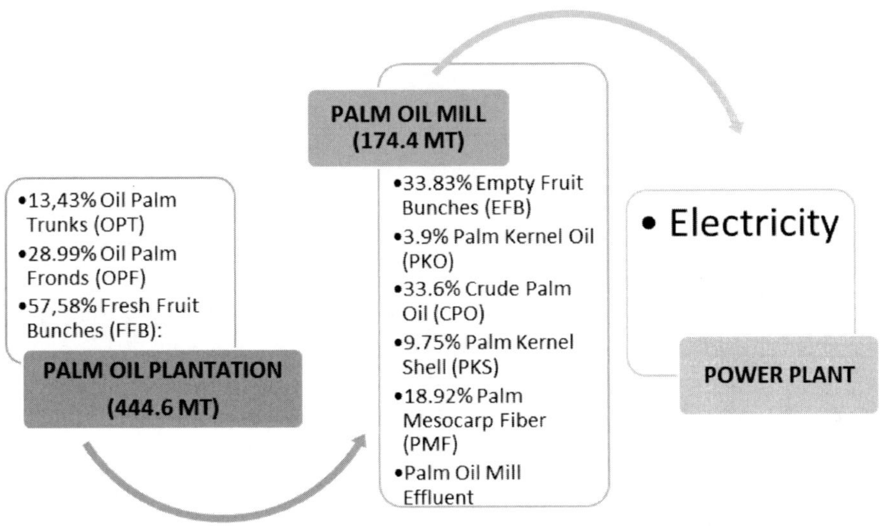

Figure 5. Oil palm plantation waste mass in Indonesia in 2020 based on Aprianti et al., (2023).

One of the sectors that makes a substantial contribution to the world economy is the palm oil industry, especially in major producing nations like Malaysia and Indonesia. The production of palm oil is growing quickly due to the ongoing increase in worldwide demand. But even with its benefits to the economy, the business still confronts several obstacles, especially when it comes to handling the garbage it produces. Oil palm solid waste has enormous potential for use in a variety of technological advancements that might aid in the shift to a circular economy and more ecologically friendly enterprises since it is a rich source of energy and valuable components. When handled properly, this waste can reduce pollution in the environment and increase the palm oil industry's overall economic efficiency. Thus, oil palm waste and by-products can now be recycled into more beneficial and profitable products because of developments in chemical engineering and material technology research. The four primary components of empty fruit bunches - mesocarp fiber, palm kernel shells, palm fronds, palm leaves, and palm trunks - as well as their use in a variety of products, including adsorbents, catalysts, and other materials, will be covered in this book. Advances in research and technology enable the palm oil industry to adopt more sustainable practices and significantly reduce its carbon footprint. This is essential for evaluating the sustainability characteristics and environmental benefits of utilizing palm solid waste materials, as well as for understanding and quantifying their impacts (Burmana et al., 2025).

Conclusion

The palm oil company in Indonesia contributes significantly to the world economy, but it also produces vast volumes of solid waste, which can pose environmental concerns if not adequately handled. Solid waste from palm oil production includes 23% empty fruit bunches (EFB), 6.5% palm kernel shell waste (PKS), 4% palm sludge, 13% fiber, and 5% liquid waste. This garbage is still widely unused. With the correct technology, this waste can be transformed into high-value goods including bioenergy, organic fertilizers, activated charcoal, catalysts, adsorbents, and other eco-friendly materials. The sustainable use of oil palm waste not only decreases negative environmental impacts such as pollution and greenhouse gas emissions, but it also improves the economic efficiency of the palm oil industry. Oil palm trash can bring major advantages to society by producing organic fertilizers and activated charcoal, both of which are useful in a variety of sectors. With advancements in research and technology, the palm oil industry can adopt more sustainable practices, reduce its carbon footprint, and increase economic returns. Therefore, the development of innovative technologies that can transform oil palm waste into high-value products is crucial for supporting a more environmentally friendly and profitable circular economy model.

Chapter 2

A Brief Review of Biomass Processing

Abstract

Biomass, derived from plants, animals, and their residues, is a renewable resource with great potential to substitute fossil fuels and reduce greenhouse gas emissions. Its abundance and carbon-neutral nature make it an attractive candidate for sustainable energy production through various biochemical, thermochemical, and physicochemical pathways. This chapter highlights key conversion technologies include combustion, pyrolysis, gasification, liquefaction, anaerobic digestion, and fermentation, each offering distinct advantages in transforming biomass into useful energy carriers such as heat, bio-oil, syngas, biogas, and bioethanol. Despite these opportunities, biomass utilization faces several challenges, including variability in feedstock composition, high moisture content, and relatively low energy density. Therefore, the effective use of biomass requires careful selection of conversion routes and integration of suitable technologies. Optimizing these processes is essential to achieve sustainable, efficient, and economically viable biomass utilization, contributing to a cleaner and more resilient global energy system.

1. Introduction

Biomass refers to biological materials derived from plants, animals, and their associated wastes and residues. The utilization of biomass as a sustainable resource has been widely recognized, primarily due to its potential to mitigate greenhouse gas emissions. Effective biomass management should employ technologies that are sustainable, non-polluting, and efficient. Various forms of biomass - such as lignocellulosic materials, sewage sludge, and municipal solid waste - represent abundant and renewable resources that can be converted through biochemical, physicochemical, or thermochemical technologies. However, biomass also presents several drawbacks, including high moisture content, low energy density, heterogeneity, low bulk density, and the presence of contaminants (Wang et al., 2018).

Residual biomass, particularly derived from plants, can generally be categorized as materials rich in polysaccharides or lignin. These polysaccharides primarily consist of cellulose and hemicellulose, which together account for approximately 60–80% of woody biomass. Meanwhile, lignin, together with polysaccharides, forms the structural framework of plant tissues (Bonechi et al., 2017).

Globally, unexploited biomass resources are estimated to represent 10–20% of the total primary energy demand. Considering both its availability and composition, biomass emerges as one of the most promising candidates for renewable and sustainable energy production. Its universal availability allows it to serve as a substitute for conventional energy sources. The integration of biomass-based energy with conventional systems can also reduce overall energy costs. Furthermore, biomass energy is particularly attractive due to its potential to reduce greenhouse gas emissions (Bonechi et al., 2017). The release of CO_2 during biomass utilization does not contribute to an increase in atmospheric CO_2 concentration, as it originates from biogenic sources. Plants subsequently absorb this CO_2 through photosynthetic and metabolic processes, effectively closing the carbon cycle. In other words, the use of biomass accelerates the natural circulation of CO_2 in the atmosphere, which is then reutilized by plants to generate new biomass (Tursi, 2019).

In terms of availability, the Earth contains vast reserves of biomass distributed across terrestrial and marine environments. The total global biomass stock is estimated at 1.8 trillion tons on land and 4 billion tons in aquatic systems. The overall energy potential of global biomass is approximately 33,000 EJ, equivalent to more than 80 times the world's annual energy consumption (Tursi, 2019). This indicates that biomass resources are theoretically sufficient to meet global energy demands. Nevertheless, current biomass utilization remains limited - accounting for only about 14% of the world's primary energy supply, or approximately 56 million TJ per year. Moreover, its use is unevenly distributed across regions. In some developing countries, biomass contributes up to 50% of total energy consumption, primarily through the combustion of wood, shrubs, and organic waste from plants and animals. In contrast, developed nations generate only about 11% of their total energy from biomass sources (Tursi, 2019). It must also be acknowledged that the composition of biomass varies significantly depending on its source (i.e., the type of feedstock or plant species).

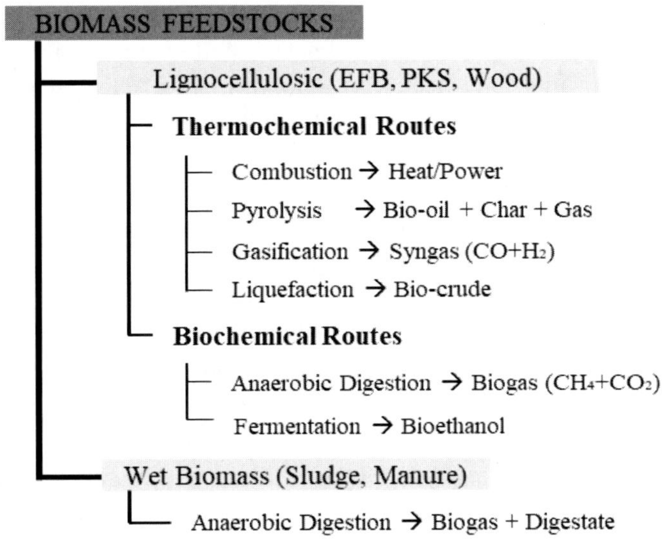

Figure 6. Overview of biomass conversion technologies.

Consequently, there are still no universally standardized guidelines for the efficient and sustainable utilization of biomass using modern technologies (Bonechi et al., 2017). In addition, biomass residues often contain various bioactive compounds that may offer higher economic value if extracted rather than combusted. These bioactive constituents - such as antioxidants, flavonoids, and lignans - have potential applications in the food, cosmetics, pharmaceutical, and materials industries. Therefore, a strategic assessment is required to determine the most appropriate pathway for biomass utilization - whether as a renewable energy source or for non-energy applications. Figure 6 provides a comprehensive overview of biomass conversion routes, categorizing technologies based on feedstock type and operating conditions. The selection of the appropriate route is crucial, as lignocellulosic biomass (EFB, PKS) is more efficiently processed through pyrolysis/gasification, while wet waste excels in anaerobic digestion.

1.1. Biomass Combustion

Roughly 90% of all renewable energy derived from biomass is generated through combustion-based processes. Such facilities are capable of utilizing a wide range of biomass feedstocks, including wood, dried leaves, hard nut shells, rice husks, and even dried animal manure (Lebaka, 2013). During combustion, biomass reacts with oxygen at elevated temperatures to produce carbon dioxide, water vapor, and heat. The amount of heat released varies depending on several parameters, particularly the type and composition of the biomass feedstock; on average, the thermal energy output is about 20 MJ per kilogram of biomass (Nussbaumer, 2003). This process represents an exothermic chemical reaction, in which the oxidation of biomass in the presence of air releases chemical energy that can subsequently be transformed into mechanical or electrical energy (Lebaka, 2013; Kaushika et al., 2016).

The combustion process takes place within specialized chambers operating at temperatures typically ranging from 800-1000°C. A crucial prerequisite for efficient combustion is that the moisture content of the biomass should remain below 50%. Conventional biomass-fired power plants - commonly utilize wood and forestry residues as fuel - can generate between 20 and 50 MWe, achieving electrical efficiencies of about 25–30%. Ongoing technological advancements, particularly the implementation of fluidized bed combustion systems and advanced gas-cleaning technologies, have significantly enhanced performance, enabling biomass-based plants to produce 50–80 MWe with electrical efficiencies reaching 30–40%.

1.2. Pyrolysis

Pyrolysis refers to the thermochemical decomposition of biomass into liquid (bio-oil), solid (charcoal), and gaseous (combustible gas) fractions through partial combustion conducted at approximately 500°C under oxygen-limited conditions (Lebaka, 2013). At elevated temperatures, the volatile compounds within the biomass are vaporized, forming gases whose vapors subsequently condense into liquid fuels through a liquefaction process. The resulting bio-oil can be stored and utilized for various heating or power generation applications. Besides liquid fuels, pyrolysis also yields charcoal, combustible gases, and a range of value-added chemical by-products, further enhancing its potential as a versatile biomass conversion technology (Kaushika et al., 2016).

As previously noted, basic pyrolysis systems typically operate at temperatures below 600°C, whereas industrial-scale pyrolysis is conducted at higher temperatures ranging from 600-1000°C, resulting in gas products with increased hydrogen content. Naturally, such high-temperature operations necessitate more advanced equipment designs and complex process management to ensure optimal performance and safety (Di Blasi, 2008).

1.3. Gasification

The gasification process involves the transformation of solid carbonaceous materials (such as solid biomass) into a gaseous mixture known as synthesis gas (syngas), which primarily consists of carbon monoxide (CO), hydrogen (H_2), and nitrogen (N_2). This syngas can serve as a fuel for electricity generation or as a feedstock in the petrochemical and refining industries for the production of various compounds, including methanol, ammonia, and synthetic gasoline (Rahimpour et al., 2012). The composition of the gas produced during gasification comprises both combustible and non-combustible components, which vary according to the type of biomass feedstock and the operational parameters of the process. Typically, the resulting gas exhibits a calorific value ranging from 4.5 to 6 MJ/m^3, equivalent to approximately 10–50% of the energy content of natural gas (Akia et al., 2014).

1.4. Liquefaction

Liquefaction is a thermochemical conversion process in which biomass is transformed into an aqueous environment under moderate temperatures (280–370°C) and high pressures (10–25 MPa). This process yields a liquid bio-crude resembling petroleum oil, along with gaseous, aqueous, and solid by-products (López Barreiro et al., 2013). The resulting liquid product possesses a high heating value and a low oxygen content, which contributes to its chemical stability as a fuel. The main objective of liquefaction is to produce an oil with a high hydrogen-to-carbon (H/C) ratio (Tursi, 2019).

Depending on the type of feedstock, two principal liquefaction routes can be identified: lignocellulosic biomass liquefaction (for dry feedstocks) and algal biomass liquefaction (for wet feedstocks). In both pathways, the raw materials must undergo pretreatment, which includes removal of contaminants from woody biomass and the formation of a stable suspension by reducing particle size - often through alkaline treatment - to facilitate efficient pumping and processing within reactors (Gollakota et al., 2018).

1.5. Anaerobic Digestion

Anaerobic digestion is a multi-stage biological process that serves the dual purpose of effective waste management and renewable energy generation. The process involves four main phases: hydrolysis, acidogenesis, acetogenesis, and methanogenesis. Throughout these stages, a series of biochemical reactions occur via microbial metabolic pathways under oxygen-free (anaerobic) conditions. These reactions decompose complex organic macromolecules into simpler compounds, ultimately producing biogas - a mixture primarily composed of methane (CH_4) and carbon dioxide (CO_2), along with small amounts of other gases - and a solid by-product known as digestate. Common feedstocks utilized in anaerobic digestion include sewage sludge, agricultural residues, municipal solid waste (MSW), and animal manure, all of which provide suitable substrates for microbial degradation and biogas production (Tursi, 2019).

1.6. Fermentation

The fermentation of organic materials involves a sequence of biochemical reactions in which simple sugars (such as hexoses and pentoses) are converted into ethanol and carbon dioxide (CO_2) under anaerobic conditions through the metabolic activity of microorganisms, primarily yeasts. The microorganism most commonly employed for this process is Saccharomyces cerevisiae, while the feedstocks utilized are typically classified into three main categories: sugars, starch-based materials, and lignocellulosic substrates.

The theoretical conversion yield of fermentation corresponds to approximately 51.14 g of ethanol and 48.86 g of CO_2 per 100 g of hexose or pentose substrate. Alongside ethanol and CO_2, small amounts of glycerol and carboxylic acids are also generated as secondary by-products. The efficiency and quality of the fermentation process are influenced by several key parameters, including the type of feedstock, temperature, pH, inoculum concentration, and fermentation duration (Strezov et al., 2014).

Conclusion

Biomass - organic materials from plants, animals, and residues - offers a promising, abundant, carbon-neutral renewable resource to replace fossil fuels and cut greenhouse gas emissions via biochemical, thermochemical, and physicochemical conversion. The global biomass reserve, estimated at over 1.8 trillion tons on land and 4 billion tons in aquatic systems, possesses an energy potential exceeding 33,000 EJ - sufficient to meet worldwide energy demands multiple times over. However, current utilization remains limited to approximately 14% of global primary energy, with marked regional disparities. Several conversion technologies have been developed to optimize biomass utilization. Combustion, accounting for nearly 90% of biomass-derived energy, directly converts biomass into heat and power. Pyrolysis thermochemically decomposes biomass into bio-oil, charcoal, and gases under oxygen-limited conditions, while gasification produces synthesis gas (CO, H_2, and N_2) for use as fuel or chemical feedstock. Liquefaction converts biomass into bio-crude under moderate temperatures and high pressures, offering a stable, high-energy liquid fuel alternative. In contrast, anaerobic digestion and fermentation employ biological pathways - producing methane-rich biogas and ethanol, respectively - from organic substrates. Despite its advantages, biomass utilization faces challenges such as variable composition, high moisture content, and low energy density. Moreover, biomass residues often contain valuable bioactive compounds with potential non-energy applications. Therefore, strategic optimization and integration of conversion technologies are crucial for ensuring the efficient, sustainable, and economically viable use of biomass in future energy systems.

Chapter 3

The Potential of Oil Palm Empty Fruit Bunches

Abstract

The use of Empty Fruit Bunches (EFB) from oil palm has become an increasingly essential waste management approach in the palm oil industry. EFB, which makes up a large amount of palm oil manufacturing waste, is high in cellulose and thus a valuable raw material for a variety of uses. Initially discarded or used as fuel, EFB is now being explored for its potential in producing renewable energy, organic fertilizers, biodegradable products, and as a reinforcement in composite materials. The high cellulose content in EFB offers numerous benefits, including its use in the production of cellulose derivatives, biofuels, and bioplastics. This chapter explores the different methods of utilizing EFB, including its role in biodiesel production, as a biodegradable pot, and in the creation of cellulose-based products like cellulose acetate and composites. The chapter also highlights the economic and environmental advantages of EFB utilization, particularly in the context of sustainable industry practices and resource efficiency.

1. Introduction

EFB utilization from oil palm is a strategic effort in managing the waste from the palm oil industry, which has often been considered garbage. EFB is the biomass residue formed after the palm fruit is harvested and the oil is extracted. Initially, EFB was either discarded or used as simple fuel, but with increasing awareness of environmentally friendly waste management and resource efficiency, EFB has begun to be utilized more optimally. This utilization not only reduces environmental impact but also opens up opportunities for economic value addition, whether in the form of renewable energy, animal feed, or raw materials for organic products like compost fertilizer.

Over time, various technologies and innovations have been developed to process EFB into sustainable, valuable products, turning oil palm waste into a potential new resource that supports agriculture and green industries. EFB is rich in fiber, lignin, and cellulose, making it an ideal raw material for renewable energy production through processes like pyrolysis or gasification to produce briquettes or biochar. In addition, the fiber contained in EFB can be used to make eco-friendly composite materials, replacing more expensive and less environmentally friendly synthetic materials. The utilization of EFB also includes the production of organic fertilizers, which can help improve soil fertility without negative environmental impact. When compared to the overall number of Fresh Fruit Bunches (FFB) processed, the availability of EFB is significant. EFB output averages 22% to 24% of the total weight of fresh fruit bunches processed in palm oil mills (Sawit, 2012). For every ton of fresh fruit bunches, roughly 0.21 tons (21%) of CPO are generated, 0.05 tons (0.5%) of palm kernel oil (PKO), and the remaining waste is 0.23 tons (23%) of EFB, 0.135 tons (13.5%) of fiber, and 0.055 tons (5.5%) of palm kernel shells. The byproducts of the palm oil plantation sector can all be used provided industry stakeholders manage them appropriately (Noorshashillawati, 2007).

In addition to the abundant quantity, the high cellulose content in oil palm bunches, which is around 45-50%, makes EFB a potential raw material for further processing into more useful products (Khalil et al., 2010). Furthermore, concerns about pollution from the palm oil industry and the possible organic content of palm oil waste necessitate adequate waste management by oil palm plantations. This project is part of a larger initiative to prevent negative impacts and encourage environmentally responsible industry practices. Palm oil mills currently process/utilize very little EFB. EFB is usually burned in an incinerator, discarded (open dumping), utilized as mulch in oil palm farms, or turned into compost. EFB accounts for 23% of the palm oil waste utilization as an alternative organic fertilizer, which also provides additional economic benefits (Ahmadzadeh and Zakaria, 2009).

Generally, EFB is processed into organic fertilizer because it still contains nutrients such as nitrogen, phosphorus, and potassium. However, the processing of EFB into organic fertilizer has not been optimized, leading to the accumulation of EFB in palm oil mills. Several challenges contribute to this, including the lengthy composting process, which takes about 6–12 months, the facilities required for processing, and the cost of processing the empty fruit bunches. As a result, these methods are less favored by palm oil mills. In addition to being converted into organic fertilizer, EFB has the potential to serve as a planting medium because it contains high levels of nutrients such as calcium (Ca), potassium (K), and magnesium (Mg), which can improve soil fertility (Halim et al., 2019). EFB has excellent potential as a root trainer planting medium due to its high macro and micronutrient content, which includes: C (42.8%), K_2O (2.9%), N (0.8%), P_2O_5 (0.22%), MgO (0.30%), Cu (23 ppm), and Zn (51 ppm) (Atminingsih et al., 2020). Furthermore, due to its high moisture content and lower calorific value as compared to PKS and mesocarp fibers (MF), EFB is often less recommended as a major fuel for energy use.

Based on the structure of EFB, which consists of fibers containing cellulose, hemicellulose, and lignin, it is dominated by organic fibers containing cellulose in the range of 36-42%. This cellulose level is adequate for use as a raw material in the manufacture of fibers, cellulose derivatives, microcrystalline cellulose, and nanocrystalline cellulose. As a direct application, EFB has been used to produce strong fibers, which can be utilized as fillers in rubber-based products, including car seats, mattresses, and composite boards. The physical properties of empty fruit bunches are shown in Table 1.

As a chemical source, oil palm EFB can be used to produce lignin compounds and derivatives, hemicellulose and derivatives, and cellulose and its derivatives. These compounds are obtained through an extraction process using acidic or basic solutions as solvents. Following extraction, modification is carried out through a specific process to produce derivatives of lignin, cellulose, and hemicellulose. The chemical derivatives obtained from the modification of these compounds can be used as raw materials to produce other chemicals (Sulhatun, 2005).

Table 1. Physical properties and morphology of EFB from oil palm (Sulhatun, 2005)

Parameter	Base Section of EFB	Tip Section of EFB
Fiber Length (mm)		
- Min	0.63	0.46
- Max	1.81	1.27
- Average	1.22	0.87
Fiber Diameter (mm)	15.0	114.34
Wall Thickness (mm)	3.49	3.68
Flexibility (mm)	0.54	0.49
Fiber Content (% w/w)	72.67	62.47
Non-Fiber Content (% w/w)	27.33	37.53

Table 2. Chemical composition, ultimate and proximate properties of EFB (Obada et al., 2023)

Property of EFB	Value (%) (Based on Dry Basis)
Chemical Composition	
- Cellulose	23.7-65%
- Hemicellulose	20-33.52%
- Lignin	14.1-30.45%
- Extractives	3.21-3.7%
Ultimate Properties	
- Carbon	43.80-54.76%
- Oxygen	38.29-47.76%
- Hydrogen	4.37-7.42%
- Sulfur	0.035-1.10%
- Nitrogen	0.25-1.21%
Proximate Properties	
- Moisture Content	2.40-14.28%
- Volatile Matter	70.03-83.94%
- Fixed Carbon	8.97-18.30%
- Ash Content	1.30-13.65%
- Calorific Value	17-19.5 MJ/kg

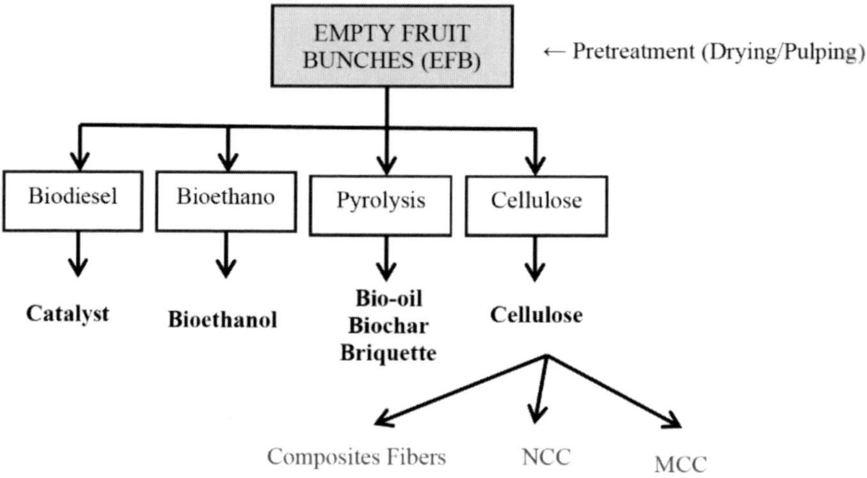

Figure 7. Schematic diagram of EFB utilization pathways.

EFB is preferred over both hardwood and softwood because much of the wood has already been utilized as raw material for the paper industry. To avoid a shortage of raw materials, EFB is more often chosen. The preference for EFB over other components of the oil palm plant, such as oil palm fronds, stems from the age of the plant. Oil palms have a long enough lifespan to continue producing fruit. If fronds were the primary raw material, the chances of obtaining them from non-productive plants would be slim. EFB, on the other hand, will be generated for as long as the palm oil business is in operation.

EFB ismade up of cellulose, hemicellulose, lignin, and a minor quantity of extractives (Table 2). Cellulose is the largest fraction, accounting for more than 40% of the dry weight of EFB, with a range of 23.7–65%. Cellulose begins to degrade at temperatures between 240–350°C, making it a significant factor in determining the thermal properties of EFB. Hemicellulose ranges from 20–40% (20.58–33.52%), while lignin is found in the range of 14.1–30.45%. Lignin is noted for its high energy content and resistance to biochemical conversion, which makes using EFB as a bioprocessing feedstock challenging. Furthermore, EFB contains a tiny quantity of extractives, approximately 3.21-3.7%, which comprise soluble chemicals including potassium, nitrogen, magnesium, calcium, phosphorus, and other minerals. The mixture of cellulose, hemicellulose, and lignin results in a complex lignocellulosic structure, making EFB difficult to convert via biochemical or physicochemical processes without appropriate pretreatment (Obada et al., 2023).

Figure 7 presents a systematic visual representation of the diverse conversion pathways for EFB derived from oil palm waste, effectively encapsulating all key processes outlined in Chapter 3 - from initial pretreatment to the generation of high-value end-products such as biofuels and cellulose derivatives.

1.1. The Utilization of EFB as a Catalyst in Biodiesel Production

Several research has been executed to use EFB as a catalyst in biodiesel production, which are listed in Table 3. EFB is commonly used as a catalyst in biodiesel production in the form of activated carbon, which is impregnated or modified with other compounds during catalyst manufacture. The synthesis methods for the catalyst vary, including wet impregnation, direct carbonization-sulfonation, and arylation using 4-benzenediazonium sulfonate

(Ibrahim et al., 2022; Wong et al., 2020). The resulting catalysts are typically solid acid-based carbon catalysts, some of which exhibit magnetic properties to facilitate separation. Important characteristics of these catalysts include a large surface area, the presence of acid sites (–SO_3H) or basic sites (via KOH doping), magnetization (if present), as well as thermal stability and the ability to be reused. The best biodiesel yield, exceeding 98%, was achieved with catalysts combining activated carbon, metals (such as Fe), and acid sites like sulfonation or chlorosulfonation, operated at optimal temperature and methanol ratios. Catalysts produced by doping KOH into EFB showed high yields with short reaction times, making them effective as basic catalysts. Meanwhile, activated carbon-based catalysts that were purely sulfonated tended to yield lower outputs but offered good stability and were more environmentally friendly.

Table 3. Studies on the utilization of EFB as a catalyst in biodiesel production

Ref.	Catalyst Synthesis Method	Catalyst Type	Catalyst Characteristics	Operating Conditions	Biodiesel Yield/ FFA Conversion
Ibrahim et al., (2022)	Wet impregnation of $Ca(NO_3)_2$ and $FeCl_3$ on EFB charcoal	Bifunctional magnetic nanocatalyst $CaO\text{-}Fe_2O_3/AC$	Surface area 106.28 m^2/g, high acid-base activity, magnetization 7.59 emu/g, high thermal stability	3 wt% catalyst, methanol: oil 18:1, 65°C, 3 hours	98.3% FAME conversion
Wong et al., (2020)	Carbonization of EFB, direct sulfonation with concentrated sulfuric acid	Solid acid carbon-based catalyst from EFB sulfonated	Amorphous structure, micropores and mesopores, total acid density 9 mmol NaOH/g	Methyl acetate: oleic acid 50:1, 10 wt% catalyst, 100°C, 8 hours	50.5% FAME yield
Lim et al., (2020)	Carbonization of EFB with acid pre-treatment, sulfonation with 4-benzenediazonium sulfonate (4-BDS)	Solid acid sulfonated carbon catalyst from EFB	Pore surface, sulfur content up to 15.98%, thermal stability, good reusability	PFAD, 20 wt% catalyst, 7 hours esterification, 110°C	98.1% yield
Tang et al., (2020)	Biomass carbonization (EFB, papaya seeds, corn), sulfonation with 4-BDS	Solid acid carbon-based sulfonated catalyst from various biomass	Surface area 639-972 m^2/g, –SO_3H groups, significant total acid density	PFAD, 5 wt% catalyst, methanol: PFAD 30:1, 100°C, 4 hours	72.1% highest FAME yield for corn catalyst (best)

Ref.	Catalyst Synthesis Method	Catalyst Type	Catalyst Characteristics	Operating Conditions	Biodiesel Yield/ FFA Conversion
Ibrahim et al., (2019)	Chemical activation, $FeCl_3$ impregnation, chlorosulfonation	Magnetic acid catalyst based on EFB (AC-Fe(x)-SO_3Cl)	Surface area 20.42 m^2/g, high acidity 29,520 μmol/g, magnetization 7.67 emu/g	4 wt% catalyst, methanol: PFAD 16:1, 100°C, 3 hours	98.6% conversion
Yaakob et al., (2012)	EFB combustion, doping with KOH	Heterogeneous solid ash-based KOH-doped EFB catalyst	High K composition, high basicity, adequate carbon content, XRD confirms potassium compound	15-20 wt% catalyst, methanol: oil 15:1, 65°C, 45-90 minutes	>98% oil conversion to biodiesel

Notes:

- EFB: Empty Fruit Bunch (Oil Palm Empty Fruit Bunch)
- AC: Activated Carbon
- FFA: Free Fatty Acid
- FAME: Fatty Acid Methyl Ester (Biodiesel)
- PFAD: Palm Fatty Acid Distillate.

1.2. The Utilization of Empty Fruit Bunches (EFB) as a Biodegradable Pot

In addition to its use as a catalyst, EFB can also be utilized as a planting medium. The cellulose fibers in EFB are formed into biopots to replace the use of polybags made from synthetic plastic, which contribute to environmental pollution. Cellulose, which consists of carbon elements, is in the form of fibers that can be transformed into biopolybags. These fibers possess good tensile strength and resistance to degradation. Cellulose fibers are also highly favored by microorganisms in the soil, so the biodegradation process of biopolybags made from EFB is not as prolonged as that of plastic polybags (Saragih et al., 2022).

Biopots, as seedling containers, are more environmentally friendly and support plant growth. Biopots can be directly planted with seedlings during the transplanting process, which prevents damage to or disturbance of the plant roots. Planting the seedlings with the biopot is also beneficial for seedling adaptation in the new growing environment (Lestari et al., 2022), as the main advantage of biopots is their fast biodegradation rate. The biopot is planted in the soil together with the seedling and decomposes due to environmental conditions. This allows the plant roots to penetrate the biopot walls (Barbosa and Castillo, 2020). Harahap et al., (2025) produced biodegradable pots using EFB waste and sawdust. The researchers wanted to explore how well EFB, and sawdust waste could be combined to make biodegradable pots as a long-term alternative to regular polybags for seedlings. The researchers first processed EFB through fermentation to reduce the content of coarse fibers and ensure that no harmful substances were present. Sawdust waste was dried and sifted to obtain fine particles. The biodegradable pots were then prepared by mixing EFB compost and sawdust waste in varying compositions, along with natural rubber latex as a binder. The physical properties of the resulting biodegradable pots were assessed, including density, moisture content, water absorption, and biodegradation. The results showed that increasing sawdust composition and decreasing EFB composition resulted in a drop in the density of the biodegradable pot, an increase in moisture content and water absorption, and a decrease in biodegradation rate. The best formulation was found to be 160 g of EFB and 20 g of sawdust waste, which resulted in biodegradable pots with a density of 0.6846 g/cm^3, moisture content of 7.09%, water absorption of 39.49%, and a biodegradation rate of 34.07%. FTIR analysis of the biodegradable pot revealed the presence of various functional groups, such as hydroxyl, carbonyl, and ether groups, which are characteristic of organic materials derived from EFB and sawdust waste. The potential for utilizing waste materials like EFB and sawdust to create a sustainable and biodegradable alternative to traditional planting media was highlighted (Harahap et al., 2025).

The use of EFB compost was also studied in combination with wet decanter solid (WDS) waste in the provision of biodegradable pots with chitosan binder. The density of the pots was measured to determine the material density of the pot ingredients. Density has an impact on the performance of seedlings grown in the product. The density of the pots generated varied greatly depending on the content of the mixture, as illustrated in Figure 8. A decrease in the EFB composition increased the density of the pots. This is due to the finer size of WDS compared to EFB. The size of a material affects the density of the pot. Smaller particle sizes in the constituent materials increase the bonding area between materials, resulting in a higher density of the pot (Lestari, 2022).

Figure 8 also shows that the greater the chitosan binder composition, the higher the resulting density. This happens due to chitosan's high amylose and amylopectin concentration, which gives it a hard texture and a strong adhesive. Therefore, the success of the entire pot-making process relies on the individual roles played by the three components: EFB, WDS, and chitosan, which together create a pot with good density. Jaya et al., (2022) found that biopots composed of OPEFB with sodium alginate had a density ranging from 0.177 to 0.264 g/cm^3. According to Jaya's research, as the content of sodium alginate grows, so does the density of the combination, as alginate works as both a matrix and an adhesive. A larger concentration of sodium alginate improves the cohesiveness of the component materials, resulting in an increase in density.

The moisture content of the pots made from the combination of EFB and WDS is shown in Figure 9. An increase in the composition of WDS and binder leads to an increase in the moisture content of the pots. This is because WDS waste has a higher moisture content compared to EFB. Materials with high moisture content will naturally have higher water content. High moisture content in organic pots generally leads to increased susceptibility to damage, both due to microbiological deterioration and chemical reactions. Furthermore, moisture content influences the acceleration of changes and the decomposition of organic materials in the growing medium and organic pots. The optimal composition for planting media based on moisture content is at an EFB:WDS ratio of 70:30 with the addition of 10% chitosan, which results in a moisture content of 10.62%.

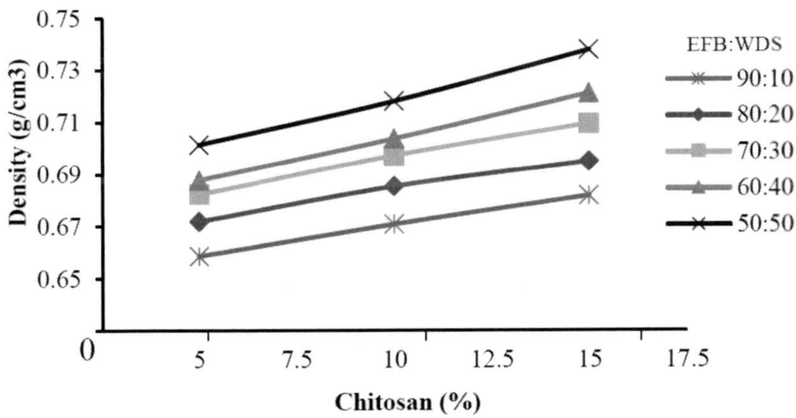

Figure 8. Effect of EFB: WDS composition and chitosan concentration on pot density.

The biodegradability test of the pots is shown in Figure 10. The highest biodegradation value of the pots was obtained with an EFB: WDS composition of 90:10 and a chitosan binder concentration of 5%, yielding a biodegradation rate of 45.05%. The EFB content contains nutrients such as NPK, which enhances its degradation ability. Therefore, the EFB content significantly influences the biodegradation of the planting medium. The biodegradation test results indicate a decrease in biodegradation with an increasing EFB to WDS ratio. The higher the EFB composition, the greater the biodegradation observed. This occurs because EFB compost has properties that allow it to decompose easily and is relatively homogeneous, which can enhance plant growth (Nirmala et al., 2020). The easily degradable nature of EFB leads to an increase in biodegradation values in the product as the EFB content in the planting medium increases.

Adding chitosan to the planting medium improves its resistance to degradation due to its antibacterial qualities and ability to create strong hydrogen bonds with EFB compost and WDS. Increasing the chitosan concentration strengthens the planting media and reduces the likelihood of soil microorganisms degrading it. Both EFB and chitosan are degradable natural polymers that can be easily destroyed by nature and other microbes. Chitosan can be easily degraded by soil microorganisms when combined with EFB and WDS. The addition of chitosan as a binder proved to be more environmentally friendly compared to synthetic plastics, which can take up to 50 years to degrade.

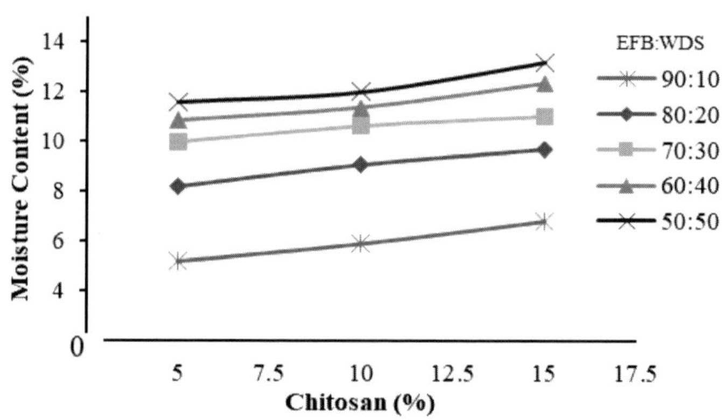

Figure 9. Effect of EFB: WDS composition and chitosan concentration on pot moisture content.

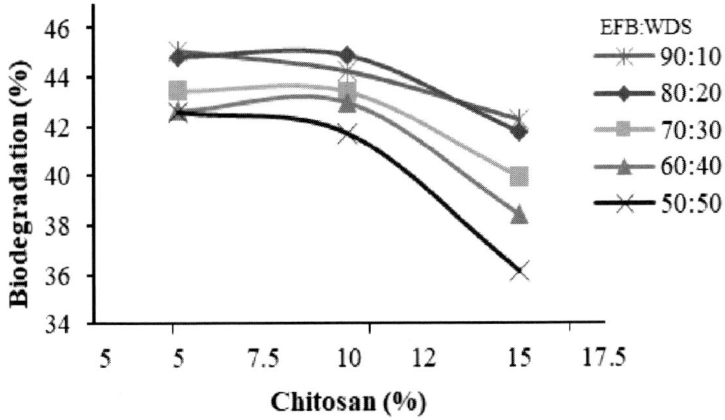

Figure 10. Effect of EFB: WDS composition and chitosan concentration on pot biodegradability.

Finally, the plant growth rate over two weeks in the pots is shown in Figure 11. Among all the tested compositions of EFB and WDS for pot making, the 70:30 EFB to WDS composition with the addition of 10% chitosan resulted in the best-quality pot. This composition yielded a pot with a pH of 6.6, which is optimal for planting, as the ideal pH range for plant growth is between 6.0 and 7.0. This composition also exhibited an optimal water absorption capacity of 8.29%. This value is considered ideal for planting media because if the absorption is too low, the growth of the plants would not be supported well, and if it is too high, the pot's durability may be compromised (Sari, 2021). The compressive strength of this formulation also showed good resistance, with a value of 0.081 MPa, which is not too high or low.

Wheatgrass seeds began to sprout on the second day, reaching a height of 1.4 cm. From day 2 to day 7, the growth of the wheatgrass increased to a length of 10.5 cm by day 7. From day 8 to day 14, the growth was less significant, with a final height of 11.4 cm on day 14. Figure 11. also compares the results with a conventional polybag, showing that by day 14, the wheatgrass in the polybag reached a height of 13.4 cm. The growth rate in the polybag was faster, as it included a planting medium made of branded fertilizer soil, which has a porous nature that facilitates water absorption and increases soil fertility. This made the growth rate in the polybag more favorable than in the planting medium, but the difference was not significant.

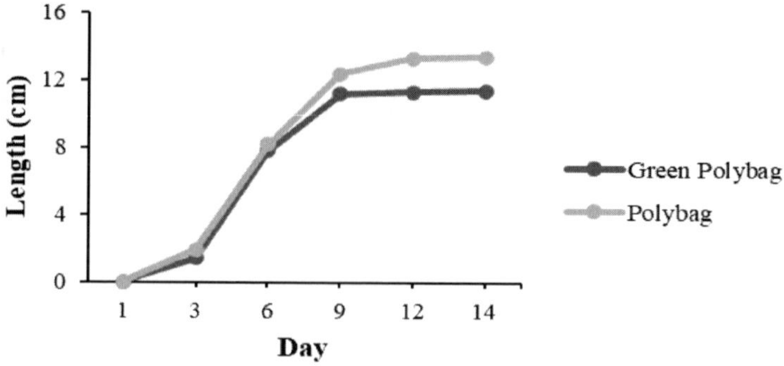

Figure 11. Comparison of growth rate in EFB-WDS pots and polybags.

From an environmental and practical perspective, the planting medium is more environmentally friendly and practical because it can be directly planted into the soil without the need to remove the pot, unlike plastic containers. Additionally, the planting medium decomposes quickly and does not cause environmental damage. EFB compost is highly beneficial for plant growth. Composted EFB contains nutrients such as water 52%, carbon 35%, potassium 5.53%, nitrogen 2.34%, phosphorus 1.14%, calcium 1.15%, and magnesium 0.96% (Satria et al., 2015). The results of the wheatgrass seed planting, using both green polybag and conventional polybag, can be seen in Figure 12.

(a) (b)

Figure 12. Wheatgrass growth results: (a) in pots; (b) in polybags.

1.3. The Utilization of Empty Fruit Bunches (EFB) in Bioethanol Production

Yoricyad et al., (2016) also investigated the use of EFB as a raw material for bioethanol synthesis using delignification and hydrolysis procedures. EFB, which is an agricultural waste with high lignocellulosic content, requires an initial step to remove lignin through delignification. This study explored two delignification methods, using NaOH with and without the addition of an ionic liquid. The results showed that the use of choline chloride (ChCl) greatly increased the yield of cellulose (80.66-89.62%) and hemicellulose (40.06-44.51%) when compared to the technique without ChCl. The lignin content was also greatly reduced, from 18.99% without ChCl to just 3.62% with the addition of 15% ChCl. Overall, the delignification rate increased by up to 61.45% using the ChCl system.

The next stage was the hydrolysis of the delignified cellulose using a combination of ChCl and sulfuric acid as a catalyst. The best results were obtained under conditions of 20% ChCl with a reaction time of 90 minutes, yielding a maximum glucose concentration of 37.96%. This value was 4.09% higher compared to hydrolysis without the addition of ChCl. This study demonstrates the effectiveness of the ionic liquid ChCl in both stages, delignification, and hydrolysis. Its use not only increases the yield of cellulose and hemicellulose but also significantly reduces the lignin content. This affirms the potential of ionic liquids as solvents and catalysts in the lignocellulosic biomass biorefinery, particularly for converting agricultural waste like EFB into fermentable sugars that can be utilized in bioethanol production (Yoricyad et al., 2016).

1.4. The Drying of Empty Fruit Bunches (EFB)

Empty Fruit Bunches (EFB), as a biomass source from oil palm plantations, is typically burned or used as soil mulch. Both methods are considered environmentally unfriendly and economically inefficient. EFB fibers are known for their strength and stability, and they can be easily processed into various sizes for applications such as mattress and cushion production, soil stabilization for erosion control, landscaping and horticulture, ceramic and brick manufacturing, and fiberboard production (Hasibuan & Daud, 2004).

Traditionally, EFB fibers are dried using exhaust gas in a diesel-powered rotary dryer. However, this approach might impair the fiber quality by producing color changes to brown and tangling. Furthermore, the usage of rotary dryers increases the risk of dust explosions since dust is close to the diesel flame inside the dryer. This issue can be resolved by replacing exhaust gas with superheated steam as a heating medium. The use of superheated steam prevents fiber browning, as industries prefer fibers with bright colors. Additionally, drying EFB fibers using superheated steam minimizes dust explosions, as superheated steam lacks oxygen (Hasibuan & Daud, 2009).

Hasibuan & Daud (2009) conducted research on drying EFB fibers using superheated steam. In this study, the rotary dryer was replaced with a trough-type belt dryer to avoid fiber tangling that occurs when using a rotary dryer. The results showed that fibers dried with superheated steam were lighter compared to those dried with hot air or undried fibers. Additionally, the EFB fibers were brighter, with a higher hue angle and a lower browning index. In terms of mechanical properties, drying with superheated steam produced EFB fibers that were stronger and more elastic, particularly when the steam temperature was set at 170°C. Lastly, a scanning electron microscope (SEM) analysis of the fiber's microstructure showed that drying with superheated steam (especially with a flow rate of 0.40 m/s) effectively removed silica particles from the fiber surface without damaging the fiber structure. This phenomenon opens the possibility of a new process for drying EFB fibers while removing silica before the pulping process for paper production, as the presence of silica can damage the pulping equipment. However, if the temperature used is too high (>200°C) or the steam flow rate exceeds 0.44 m/s, the EFB fiber structure may crack and degrade, reducing the mechanical strength of the fibers.

Burhani et al., (2022) evaluated the effectiveness of freeze-drying, spray-drying, and oven-drying on EFB CNF suspensions. Freeze-drying of CNF resulted in a white, light, and easily moldable material with a density of 0.01-0.02 g/cm^3 and a porosity of around 98%. This CNF is highly suitable for use as an adsorbent. During spray-drying, the CNF generated irregular particles with an average diameter of 2-5 μm, similar to spheres. Spray-drying produces CNF particles, which can be employed as a reinforcing filler in composite materials, including controlled medication delivery systems. Finally, oven drying resulted in CNF with a thin, rigid film-like shape and a rough surface, indicating a loose arrangement of CNF. The resulting dry CNF consisted of layers showing agglomeration, where the nanofibers were no longer present. Therefore, oven drying is not recommended as a drying method for nanocellulose.

Al-Kayiem and Yunus (2013) compared the drying process of EFB using four methods: (1) open solar drying; (2) combination of direct and indirect solar drying; (3) drying with backup heat source; and (4) hybrid dryer. The study found that combined solar drying took 52-80 hours, whereas open solar drying took 100 hours. Drying with a backup heat source accelerated the process to 48–56 hours. Finally, the hybrid dryer required the shortest time, 24–32 hours. These findings demonstrate that the combination of solar energy and backup heat sources (hybrid drying) effectively enhances drying performance and provides a practical method for producing solid fuel from EFB.

1.5. The Pyrolysis of Empty Fruit Bunches (EFB)

Handoko et al., (2021) investigated the possible conversion of EFB into biochar, bio-oil, and syngas using batch pyrolysis at atmospheric pressure, 600°C, and 60°C/min. Proximate analysis found that the biochar produced contained 6.24% moisture, 26.30% ash, 14.14% volatile matter, and 53.32% fixed carbon. Meanwhile, ultimate analysis of the biochar showed compositions of C (55.76%), H (2.92%), O (12.5%), S (0.34%), and N (2.18%). The calorific value of the biochar was 4,966 kcal/kg adb, resembling the characteristics of sub-bituminous coal. These findings suggest that biochar has the potential to serve as a coal substitute to reduce CO_2 emissions in thermal power plants. Furthermore, the fractionation of bio-oil revealed that the initial boiling point (IBP) was 67.8°C, while the final boiling point (FBP) was 666.8°C. Kerosene (36.2%) and diesel made up the majority of the bio-oil, demonstrating its potential as a liquid fuel following further processing. The non-condensable gas products included CH_4 (13-17% vol), H_2 (28-33% vol), CO (17-26% vol), and CO_2 (16-31% vol), with calorific values ranging from 2,600-3,300 kcal/Nm^3. This composition indicates the potential for utilizing syngas as a source of energy for pyrolysis or for chemical synthesis such as methanol and dimethyl ether.

Rahayu et al., (2021) conducted more thorough pyrolysis research of EFB at temperatures ranging from 350°C to 550°C. Pyrolysis was performed at a heating rate of 10°C/min with EFB particle size <2 mm. Initial characterization showed that EFB had 9.13% moisture, 9.33% ash, 89.39% volatile matter, and 17.48% fixed carbon. This study showed that maximal biochar formation was possible at low temperatures and heating rates, but maximum bio-oil production occurred at high temperatures and heating rates.

For example, the maximum yield (36.92%) was achieved at a process temperature of 350°C. The maximum bio-oil output (46.60%) was achieved at a process temperature of 550°C.

In addition to operational circumstances, pyrolysis research of EFB has concentrated on determining kinetic parameters for the pyrolysis process. Kinetic parameters are critical for designing appropriate gasification equipment since they are based on kinetic studies conducted during the pyrolysis process. Without these kinetic parameters, process improvements and biomass gasification integration may be impossible, resulting in production instability and inefficiency (Sidek et al., 2022). Sidek et al., (2022) investigated and compared the kinetic parameters of EFB pyrolysis using both model-free and model-fitting techniques. The breakdown rate of EFB was determined using data acquired from TGA analysis. The model-fitting strategy in the study used the Coats-Redfern method, whereas the model-free approach used the Friedman and Flynn-Wall-Ozawa (FWO) methods. Both modeling approaches (model-fitting and model-free) were used to predict the activation energy (Ea) and pre-exponential factor (A) as kinetic parameters for EFB pyrolysis at heating rates of 10, 20, and 50 K/min. The Coats-Redfern approach resulted in a regression coefficient (R^2) of 0.9911, with Ea and A values of 76.53 kJ/mol and 16,269.44 min^{-1}. In the model-free approach, the Friedman method resulted in the highest R^2 value of 0.9987, with a conversion factor of 0.6, Ea of 230.22 kJ/mol, and A of 7.53×10^{19} min^{-1}. The FWO method yielded an R^2 of 0.9988, a conversion factor of 0.7, Ea of 473.37 kJ/mol, and A of 2.60×10^{19} min^{-1}. Based on the regression coefficient values, the model-free methods (Friedman and FWO) were considered more suitable for representing the pyrolysis kinetics of EFB. Industrial biomass conversion could use these methodologies, particularly during the pyrolysis process, because calculating kinetic parameters is critical for equipment design, optimization, and scale-up development.

1.6. Cellulose-Based Products from Empty Fruit Bunches (EFB)

Cellulose is the primary component of EFB, accounting for 42.7–65% of its weight (Hossain et al., 2016). Cellulose is strongly linked to hemicellulose and lignin. The extraction of cellulose necessitates extensive chemical treatments. The glucose residues establish a 180° bond with one another, resulting in repeating units in the cellulose chain made up of two cellulose molecules that constitute a cellobiose unit. Cellulose is one of the most common polymers in nature. While cellulose is found in pure form in cotton fibers, the majority of it is mixed with lignin and other polysaccharides, such as hemicellulose, in the cell walls of woody plants, including softwood and hardwood, straw, and bamboo. Cellulose consists of anhydroglucopyranose units linked by β-1,4-glycosidic linkages, forming an unbranched macromolecular chain (Hon and Shiraishi 2001). Each anhydroglucopyranose unit has three hydroxyl groups, as seen in Figure 13. Cellulose's empirical formula is $(C_6H_{10}O_5)n$, with n ~ 1500 and a molecular weight of ~ 243,000 (Sun et al., 1999).

Cellulose contains approximately 62–87.6% crystalline regions, with the remaining being amorphous (Hall et al., 2010). Pure cellulose is almost never found in nature, as it is always bound with other materials such as lignin and hemicellulose. The cellulose molecules form microfibrils of glucose units that are linked together, forming long polymer chains. The presence of lignin and hemicellulose around cellulose acts as a major barrier to hydrolyzing cellulose.

Cellulose properties include both physical and chemical characteristics. Long-chain cellulose has stronger physical properties, making it more durable against degradation from heat, chemicals, and biological influences. Other physical properties of cellulose include (Vendantu, 2025):

1. It can degrade through hydrolysis, oxidation, or both chemically and mechanically, which leads to a decrease in its molecular weight.
2. It's insoluble in water and organic solvents, but somewhat soluble in alkaline solutions.
3. In its dry state, cellulose is hygroscopic, hard, and brittle. However, when cellulose contains more water, it becomes soft.
4. Cellulose in crystalline form has better strength compared to its amorphous form.

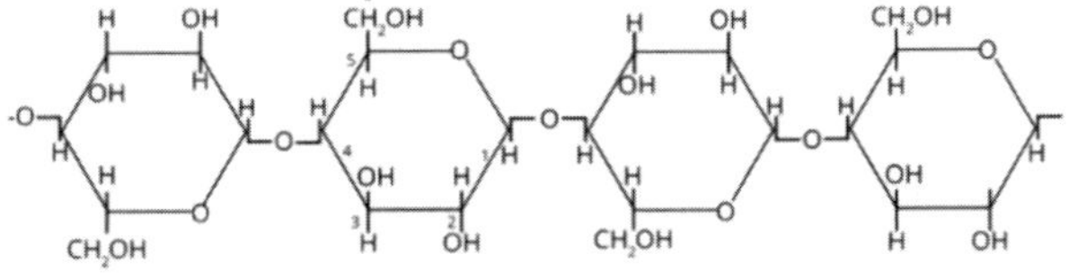

Figure 13. Structure of cellulose (Sun et al., 1999).

Ethyl cellulose, methyl cellulose, carboxymethyl cellulose, and other cellulose derivatives are commonly utilized in pharmaceutical formulations for oral, topical, and injectable usage. Seprafilm™, a surgical adhesion prevention product, has carboxymethyl cellulose as a major component. Recently, the use of microcrystalline cellulose in emulsions and semi-solid injectable formulations has been debated. These types of cellulose are used in formulations because they are innocuous and have great biocompatibility with humans (Lukova et al., 2023).

Cellulose consists of two parts: amorphous and crystalline. According to Srivasta et al., (2022), cellulose can be found in crystalline microfibrils in forms such as cellulose I, II, III, and IV. The crystalline fraction is expressed as a percentage known as the crystallinity index. The determination of cellulose structure can be carried out through X-ray diffraction, NMR, and FTIR. Cellulose I is the original form of cellulose, consisting of two allomorph crystals: Iα and Iβ. According to electron diffraction tests, cellulose Iα has a triclinic unit cell, while cellulose Iβ has two monoclinic unit cells, both arranged in parallel chains, with different ratios in a fiber depending on its origin. Cellulose Iα is abundant in bacterial cellulose and valonia, while Iβ is found in cotton and wood cellulose.

In addition to cellulose I, there is cellulose II, which is formed by precipitating cellulose from a solution into water at room temperature or slightly higher during the technical process of spinning synthetic cellulose fibers. Cellulose II is also obtained through the mercerization of cotton, which involves the formation of sodium cellulose through the interaction of the polymer with sodium hydroxide and subsequent neutralization or removal of sodium hydroxide. The change of cellulose I into cellulose II is often irreversible, while some argue that sodium cellulose can be partially converted back to cellulose I. The hydrogen bond system in cellulose II is more complex than in cellulose I, resulting in a greater number of intermolecular cross-links. Cellulose III is produced by treating cellulose I or II with ammonia at -30°C and then recrystallizing the sample by evaporating the ammonia. Based on the degree of polymerization and solubility in 17.5% sodium hydroxide (NaOH), cellulose can be classified into three categories (Kamide, 2005).

1. α-cellulose is a long-chain cellulose that is resistant and insoluble in 17.5% NaOH or strong alkaline solutions, with a polymerization degree of 600-15,000. α-cellulose is an indication of cellulose purity. High-purity cellulose (>92%) fulfills explosive production criteria. Higher α-cellulose concentration indicates better material quality.

2. β-cellulose, a short-chain cellulose, is soluble in 17.5% NaOH or strong alkaline solutions and has a polymerization degree of 15-90. Neutralizing the extract can cause β-cellulose to precipitate.
3. γ-cellulose is a short-chain cellulose that is soluble in 17.5% NaOH or strong alkaline solutions and has less than 15% polymerization.

Cellulose is naturally formed as individual molecules (linear chains of glucosyl residues), with approximately 30 individual cellulose molecules combining to create larger units known as protofibrils. These particles join to form larger units called microfibrils. These microfibrils eventually mix to produce cellulose fibers. Cellulose fibers have fine characteristics, are linear in structure, and possess high intramolecular and intermolecular hydrogen bonding. As a result, cellulose is not thermoplastic and is difficult to break down without the assistance of chemicals or enzymes.

In nature, cellulose fibers are not purely crystalline but contain regions that are amorphous. The total surface area of cellulose is larger than that of other fibers of the same dimension. According to Rebenfeld (1991), the heterogeneous structure of fibers means that some small portions of the fiber become hydrated when soaked in solutions, and some micro-pores can be penetrated by larger molecules, such as enzymes. The natural crystalline structure of cellulose is where all atoms have fixed positions relative to each other, and the crystalline aggregates form a strong molecular component to prevent the penetration of molecules such as enzymes and water.

Strong hydrogen bonds along the cellulose chain help to sustain its structure. In natural plant cellulose, cellulose chains are linked together to create highly crystalline microfibrils, with hydrogen bonds connecting each cellulose chain. Cellulose is isolated from plant cell walls, woody materials, cotton seed hairs, bark, and marine plants. The amount of cellulose in fibers varies according to the source and is generally related to materials such as water, waxes, pectin, proteins, lignin, and mineral substances. Cellulose obtained from wood requires an extensive process to remove hemicellulose and lignin.

The pulping process involves separating cellulose from lignin and hemicellulose. Pulp production can be carried out mechanically, semi-chemically, or chemically (Henriksson et al., 2024). The mechanical method is carried out through the groundwood process, where blocks of wood are pressed with a wet and rough grinding stone rotating at speeds of 1000–1200 m/min. Semi-chemical pulping combines chemical and mechanical processes. In this procedure, chemicals are used to soften wood, and the resulting pulp is mechanically processed. Chemically, pulping can be done using sulfite, alkaline, and sulfate processes to dissolve lignin and hemicellulose, leaving cellulose as a solid. The sulfite process uses a mixture of sulfuric acid (H_2SO_3) and bisulfite ions (HSO_3-) to dissolve lignin as lignosulfonic acid that is soluble in the cooking solution. The alkaline process involves cooking wood with NaOH solution. The sulfate (Kraft) process uses NaOH and Na_2S solutions. The use of these two substances enhances delignification and pulp strength. Another delignification method is steam explosion (Jacquet et al., 2015), in which wood pieces are subjected to high pressure and temperature using an autoclave.

1.6.1. Cellulose Acetate Products from Empty Fruit Bunches (EFB)

Several studies have been conducted to produce cellulose from EFB, including those by Purba et al., (2023) and Gaol et al., (2013). Gaol et al., (2013) produced cellulose acetate from α-cellulose extracted from EFB. EFB has a high α-cellulose content (45.8%), making it a viable raw material for cellulose acetate synthesis. Cellulose acetate is an important organic ester with various applications in the textile, photographic, membrane industries, and other fields. The main processes in the process include activation, acetylation, hydrolysis, neutralization, and drying.

The activation step involved adding glacial acetic acid to the α-cellulose and stirring for 3 hours. Then, acetic anhydride was used as an acetylating agent, and the acetylation process was carried out for 2, 2.5, 3, and 3.5 hours. Water and glacial acetic acid were added during the hydrolysis step, and the reaction was allowed to run for 30 minutes. After that, sodium acetate was added for 5 minutes of neutralization. The finished product was washed to remove the acetic acid odor and dried at temperatures less than 50°C.

This study discovered that the best reaction time for acetylation of cellulose from EFB was 2.5 to 3 hours, as the maximum degree of substitution (DS) and acetyl content were produced during this period. The produced cellulose acetate had a DS value between 1.387 and 1.680, classifying it as monoacetate cellulose, with acetyl content varying between 27.092% and 31.080% depending on the acetylation time. However, when the acetylation time exceeded 3 hours, cellulose degradation occurred, leading to a reduction in DS and acetyl content. Furthermore, the melting temperature of the produced cellulose acetate ranged from 160 to 200°C, with the higher the acetyl content and DS, the higher the melting point due to the superior thermal resistance of cellulose acetate over cellulose. FTIR analysis validated the acetylation reaction's success by revealing typical acetyl group peaks. This study successfully achieved the production of cellulose acetate from EFB using the cellulase process, characterized its properties, and demonstrated the significant potential of EFB waste as a value-added product (Gaol et al., 2013).

Additionally, Purba et al., (2023) also investigated the same process by producing a biofilm from cellulose acetate (CA) derived from EFB and corn starch (CS) as potential materials for environmentally friendly bioplastics. The background of this study lies in the environmental problems caused by the widespread use of conventional plastic bags, which are non-biodegradable and cause soil pollution. EFB was introduced as a promising source of cellulose for producing CA, a natural polymer with good mechanical properties suitable for bioplastic production. The process involved isolating cellulose from EFB, synthesizing CA, and mixing the solution to produce the final biofilm. EFB fibers were dried, cleaned, and cut, followed by cellulose isolation through delignification using 12% NaOH at high temperature, then bleaching with H_2O_2. The resultant cellulose was activated and acetylated using glacial acetic acid and acetic anhydride, with sulfuric acid acting as a catalyst. After the reaction, the precipitate was filtered and dried. For biofilm production, $CaCO_3$ and corn starch were dissolved, mixed with cellulose acetate, sorbitol, and charcoal, and heated. The resulting solution was cast on wax plates and dried to form biofilms with a thickness of 0.2-0.4 mm, then peeled off after further drying. Characterization techniques such as FTIR and XRD were used to analyze the properties of cellulose and CA. The following chemistry equation shows how cellulose acetate is formed:

$$\underset{\text{cellulose anhydrite}}{C_6H_7O_2((OH)_3)_x(s)} + \underset{\text{cellulose}}{3_x(CH_3CO)_2O(l)} \rightarrow \underset{\text{cellulose acetate}}{C_6H_7O_2((OCOCH_3)_{3x}(s)} + \underset{\text{acetic acid}}{3_xCH_3COOH(l)}$$

FTIR analysis of cellulose acetate (CA) showed the presence of hydroxyl (O-H) and ester (C-O) groups, indicating the richness of cellulose content. The cellulose obtained from EFB fibers had a cellulose content of 39.42 wt%, with 80.0 wt% of it being α-cellulose. XRD analysis revealed that CA had an amorphous crystalline structure with a crystallite size of 4.99 nm. The mechanical properties of the biofilm showed significant variation, with tensile strength ranging from 1.461 MPa to 8.442 MPa and elongation between 2.611% to 8.808%. The biofilm with a composition of 2 g CA, 0.5 g $CaCO_3$, and 5.5 g corn starch (CS) exhibited the maximum tensile strength and elongation (8.442 MPa and 8.808%, respectively), while the biofilm with 5 g CA, 0.5 g $CaCO_3$, and 3.5 g CS performed the worst. This variability was caused by agglomeration and uneven distribution of the CA filler within the CS matrix (Purba et al., 2023).

Bio-polybags made from biofilm have the potential to replace ordinary plastic bags, minimize environmental pollution, and deliver nutrients to the soil. Furthermore, these bio-polybags are more efficient, practical, and closely resemble traditional plastic bags, compared to plant pot-based bio-polybags which are thicker and more rigid. The successful utilization of EFB, a significant agricultural waste, to produce CA-based biofilms with desired mechanical properties demonstrates this approach's potential. Not only does it address environmental issues related to plastic bags, but it also provides a valuable way to recycle waste material, contributing to the development of more sustainable agricultural systems.

Cellulose acetate is also one of the most widely used polymers in membrane production, generally processed through the phase inversion method. This technique involves converting a homogenous polymer solution into a solid membrane via a controlled phase transition from liquid to solid. The process typically begins by casting the polymer solution onto a flat substrate, followed by immersion in a coagulation bath, usually using distilled water. This immersion promotes the exchange of solvents and non-solvents, resulting in the removal of the solvent from the polymer matrix and the development of a thin membrane layer. To make an effective membrane, the polymer must be thoroughly dissolved in the solvent of choice. Additionally, several parameters - including the presence of additives, temperature, immersion time, polymer concentration, and solvent type - significantly affect the phase separation of kinetics and the structure of the resulting membrane.

Phase inversion is a widely applied technique for membrane fabrication, where a homogeneous polymer solution transforms from liquid phase into a solid membrane form through phase separation. In the initial stage, the polymer solution separates into two immiscible liquid phases. One of these phases, which has a higher polymer concentration, solidifies to form the base structure of the membrane. This technique allows for the formation of porous polymer membranes with a range of morphological characteristics. The type of polymer and specific additives used in the casting solution have a significant impact on the final membrane structure, which includes porosity, mechanical qualities, and surface features.

Key parameters used to evaluate membrane performance are flux and rejection. Flux is the volume of permeate produced per unit area of the membrane per unit time, while rejection measures the ability of the membrane to separate specific components from the feed stream. Several studies have reported successful cellulose acetate membrane production from EFB, with favorable physical, mechanical, and filtration performance characteristics. The abundant availability of EFB waste and the relatively lower production costs compared to conventional synthetic membranes offer significant advantages for the commercialization of EFB-based membranes. However, several challenges must be addressed, including consistency of raw material quality, process scalability, fouling control, production costs, and regulatory compliance (Yoshi et al., 2025).

1.6.2. Fiber Products From Empty Fruit Bunches (EFB)

The utilization of EFB fibers can be applied in the production of composite panels, lightweight concrete, and environmentally friendly filler materials. EFB fibers can be processed into powder, which is the result of grinding EFB fibers into fine particles of <2 mm. This powder has a relatively high calorific value, ranging from 3000 to 5000 kcal/kg, depending on its moisture content and composition (Hudayarizka et al., 2024). EFB powder, in the form of briquettes or pellets, can be utilized as an environmentally acceptable solid fuel for both residential and industrial applications. The use of EFB powder as fuel is one of the alternatives to replace coal in efforts for industrial decarbonization, while also supporting renewable biomass-based energy programs in Indonesia. Another advantage is its relatively low ash content, leading to cleaner emissions. EFB fibers have also been extensively studied as reinforcement in polymer-based, cement-based, and bioresin-based composite materials (Ayu et al., 2020; Ibrahim et al., 2020; Surya et al., 2014; Siyamak et al., 2012). Its advantageous mechanical qualities and capacity to boost composite tensile strength, elasticity modulus, and thermal resistance make it appealing for a wide range of applications, including lightweight construction materials and automotive parts.

EFB fibers are hydrophilic, so chemical treatments such as alkalization (NaOH), silanization, or other surface modifications are needed for their application in composites to improve adhesion between the fibers and the polymer matrix (Senawi et al., 2014; Rosman et al., 2014; Iskandar et al., 2024). The processing of EFB into modified fibers and powder can be carried out through several stages, including drying, shredding, sieving, and chemical/physical treatments. In the production of fibers for composites, alkalization is commonly used to reduce lignin content and improve cellulose molecular bonding. Meanwhile, for powders used as fuel or chemical feedstock, further processing such as carbonization (pyrolysis) or fermentation can also be performed depending on the intended use.

Although chemical treatments such as alkalization or silanization are often applied to enhance the bonding between natural fibers and matrices in composite materials, not all applications require such treatments. In many studies, EFB fibers and powders without chemical treatment still show potential as fillers or reinforcements in various types of composites, whether based on thermoplastic polymers, thermosets, or cement. Using untreated fibers offers several advantages:

a. Environmentally friendly, as no harmful chemicals are used
b. Lower production costs
c. Simpler processing
d. More suitable for small-scale applications or village-based household industries.

Untreated EFB fibers still contain significant amounts of lignin and hemicellulose. These components may decrease the interfacial adhesion between fibers and the matrix (such as plastic or cement). The hollow and rough structure of the fibers still provides an effective mechanical interlocking mechanism in certain composite types. Specifically, for cement- or gypsum-based composites, EFB fibers can even help reduce shrinkage cracking and improve fracture resistance, although there may be a slight decrease in compressive strength compared to fibers subjected to chemical treatments (Hanan et al., 2020; Sunardi et al., 2025; Adlie et al., 2019).

1.6.3. Microcrystalline Cellulose Products From Empty Fruit Bunches (EFB)

Microcrystalline cellulose (MCC) is a form of cellulose that has undergone partial depolymerization and purification. It is white, odorless, tasteless, in powder form, and consists of porous particles. MCC has been developed for many industrial usages such as edible films, yogurt, and especially as a reinforcement and filler in polymer matrices or biocomposites (Sainz et al., 2010; Mathew et al., 2005; Reis et al., 2014; Sekar et al., 2024).

The production of microcrystalline cellulose from EFB biomass involves isolating cellulose from EFB. The resulting cellulose is then isolated to obtain α-cellulose, which is subsequently processed to produce MCC. Several processes can be used to produce microcrystalline cellulose, including:

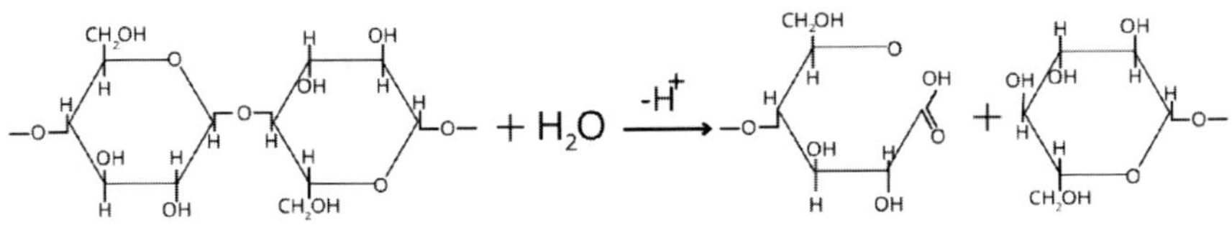

Figure 14. Acid hydrolysis reaction.

1.6.3.1. Acid Hydrolysis Process

The acid hydrolysis procedure is the standard method for manufacturing microcrystalline cellulose. According to U.S. Pat. No. 3,954,727, the hydrolysis reaction of cellulose with acid to form MCC is carried out in a batch reactor (Rasanen et al., 2010). The acid solution that can be used includes hydrochloric acid (HCl) or sulfuric acid (H_2SO_4). The acid solution dissolves the amorphous cellulose. The required operating conditions for the reaction are temperatures above 160°C. Strong acids are used to remove the amorphous part of the cellulose chain, allowing the isolation of the crystalline cellulose portion. Figure 14 shows the cellulose hydrolysis reaction with strong acid.

Numerous studies on MCC synthesis have been conducted, including those by Supian et al., (2018), Nasution et al., (2017), Nasution et al., (2020), and Samat et al., (2018). The initial cellulose content, acid concentration, hydrolysis period, and temperature all have an impact on microcrystalline cellulose hydrolysis. The amorphous sections of the cellulose chain are more easily degraded by acid than the crystalline ones. A common technique entails hydrolyzing pure cellulose with strong acid at controlled temperatures, agitation, and time intervals. The chemical process starts by breaking the polysaccharide bonds at the surface of the cellulose fibers, followed by the breakdown and degradation of the amorphous regions, releasing the crystalline cellulose. Following hydrolysis, the suspension is diluted with water and washed numerous times. The type of mineral acid employed in the hydrolysis step influences the surface characteristics of the cellulose crystals.

1.6.3.2. Steam Contact Process

The second process for producing microcrystalline cellulose (MCC) is described in U.S. Pat. No. 5,769,934, where cellulose is contacted with pressurized steam at temperatures around 180°C and 350°C for a sufficient amount of time to achieve the Levelling-Off Degree of Polymerization (LODP) (Ha and Landi, 1997). The steam contact process aims to hydrolyze cellulose and remove lignin and hemicellulose. Saturated steam is continuously fed into the reactor until a pressure of 430 psig is reached. The reactor pressure ranges from 390 psig (2.689 Pa) to 430 psig (2.965 Pa). The advantage of this process is that it does not require acid solutions, and the MCC produced is in a colloidal form.

1.6.3.3. Gas Hydrolysis Process

The gas hydrolysis process involves the use of gas for hydrolysis, as described in U.S. Pat. No. 5,543,511 (Nguyen, 2002). Cellulose is partially hydrolyzed inside a pressurized reactor using water while maintaining the reactor temperature at the reaction temperature. Oxygen or carbon dioxide gas is injected into the reactor, with pressure maintained between 0.1 to 60 bar at 20°C. The ratio of cellulose to water in the reactor is 1:8 to 1:20 (V/V). The advantage of this process is that it can yield above 95%. This process is considered environmentally friendly because the waste water produced no longer contains inorganic salts. However, this process is only suitable for pure cellulose feedstock (Khalil et al., 2010).

1.6.3.4. Reactive Extraction Process

The production of microcrystalline cellulose using the reactive extraction process is outlined in U.S. Pat. No. 6,228,213 (Karppil et al., 2010). The first extraction stage involves sodium hydroxide (NaOH) at temperatures 140°C to 170°C. It can remove lignocellulosic complex compounds. The second stage involves reacting cellulose with an acid solution at 140°C. The cellulose and acid solution are reacted in a screw conveyor. The screw rotates to apply pressure to the cellulose, mixing it with the acid solution to form microcrystalline cellulose. The advantage of this process is that it can be performed continuously to produce microcrystalline cellulose with faster reaction times and good efficiency. The particles produced are small, thus eliminating the need for additional processing to reduce particle size (Indexmundi, 2011).

1.6.3.5. Enzyme Process

The enzyme process for producing microcrystalline cellulose uses enzymes as catalysts. The enzymes used are produced by microbes, such as α-amylase, which is used for hydrolyzing starch into glucose and maltose. In cellulose hydrolysis, microbes such as *Trichoderma viride* (Ghazali et al., 2017) are employed to produce endo-cellulase enzymes that selectively break down the amorphous portion of α-cellulose. The recommended operating conditions for this process are temperatures between 50-60°C and a pH of 2.5 to 3. The advantages of this process are that enzyme hydrolysis is cleaner and more selective, operating at moderate pressure and temperature. However, enzyme hydrolysis is slower, with reaction times ranging from 24 to 48 hours. This process is advantageous in terms of reaction time, economics, and equipment.

1.6.4. Nanocrystalline Cellulose Products From Empty Fruit Bunches (EFB)

In general, there are two forms of cellulose nanomaterials derived from lignocellulosic biomass: cellulose nanofibrils (CNF) and cellulose nanocrystals (CNC), each with unique features. The simplest approach to discern between these two cellulose nanoparticles is to examine their preparation. CNF is extracted through a mechanical process, while CNC is extracted through a chemical process (Stark, 2016). CNC is an ordered crystalline structure found within cellulose fibers that consists of both amorphous and crystalline areas. CNC is generated when microfibrils break down into shorter crystalline segments. CNC is also referred to as nanowhiskers, nanorods, or rod-shaped cellulose crystals. Hydrolysis with highly concentrated acid separates the amorphous domains and breaks the local crystalline connections between nanofibrils, resulting in the formation of nanocrystalline particles. This chemical process is followed by a powerful mechanical or ultrasonic treatment. A key feature of CNC synthesized using sulfuric acid is the negative particle charge caused by the production of sulfate ester groups, which improves the stability of the nanocrystal particle phase in aqueous conditions (Novo et al., 2016). CNC deterioration is linked to the fact that hydrolysis kinetics are faster in the amorphous region than in the crystalline region due to the amorphous regions' increased permeability. Figure 15 shows an illustration of cellulose nanocrystals.

The geometric dimensions of CNC can vary greatly, with diameters ranging from 5–50 nm and lengths ranging from 100–500 nm. CNC's particular dimensions and crystallinity are determined by its cellulose supply and extraction procedures. CNC can also be used to reduce hydrophilicity and improve the integration of modified nanoparticles into hydrophobic polymer matrices. CNC has a rod-like structure and is far less flexible than CNF (Cellulose Nanofiber) due to its increased crystallinity. CNC has several benefits over cellulose fibers, including nanoscale dimensions, high specific strength and modulus, large surface area, unique optical qualities, and more.

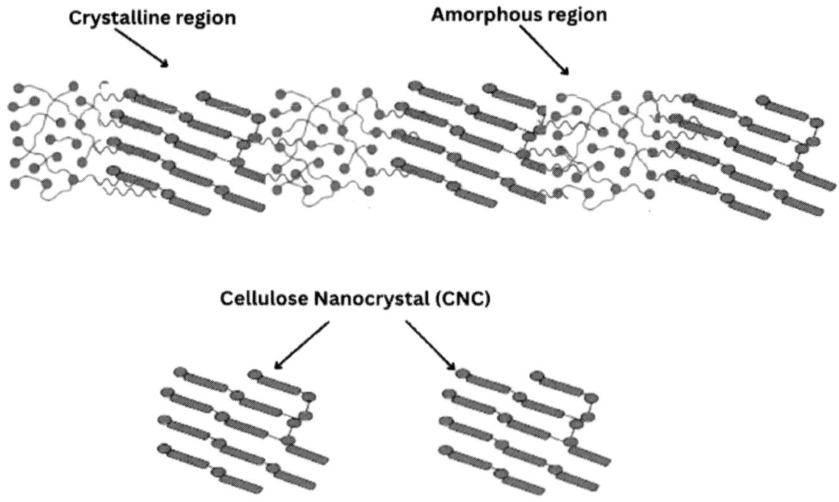

Figure 15. Illustration of the crystal structure of nanocrystalline cellulose.

1.6.5. Cellulose Applications From Empty Fruit Bunches (EFB) in Composites

The utilization of polymer materials in the industrial sector has increased dramatically. This is due to the fact that polymers are lightweight, inexpensive, resistant to corrosion, and have relatively low processing temperatures compared to metals or ceramics. In general, these polymer materials are mixed with other materials to obtain enhanced properties, which are known as composite materials.

Composite materials are created by combining two or more components on a macroscopic scale to make a third, separate substance. Different materials can also be blended on a microscopic level, such as when metals are joined. When a composite is well designed, it can provide higher quality than its separate parts. Strength, stiffness, corrosion resistance, wear resistance, attractiveness, weight, fatigue life, temperature-dependent behavior, thermal insulation, thermal conductivity, and acoustic insulation are some of the features that can be achieved using composite materials.

In general, not all of these qualities develop concurrently since they may conflict with one another. For example, thermal insulation and thermal conductivity may not coexist effectively. The primary goal of creating composite materials is to produce a new materials with specific properties for particular applications.

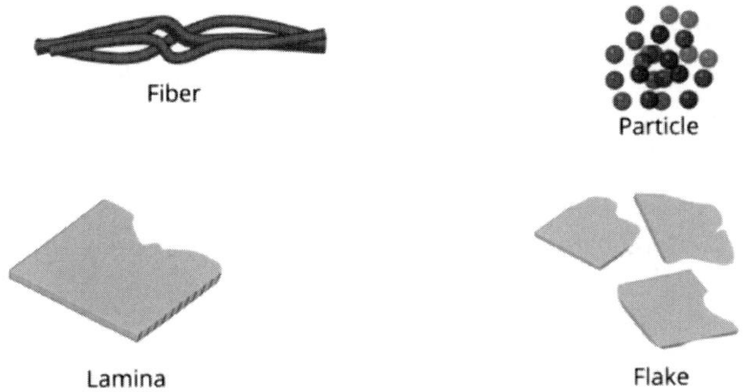

Figure 16. Different constituent forms.

Composite materials have a long history of use. While the exact first use is unknown, historical records suggest that composites have been used for various purposes. For example, straws were used to improve the strength of bricks. Plywood, which can serve as a substitute for wood, offers excellent strength and thermal resistance. Today, resin matrix composites with fiber reinforcement, which offer a high strength-to-weight ratio, have become a significant concern in the automotive and aerospace industries.

In principle, composites are formed through the combination of two or more materials, such as metals, polymers, or ceramics. While the combinations of materials are almost limitless, the constituent forms are more limited. Fibers, particles, laminates (layers), flakes, fillers, and matrices are some of the most common composite constituents. Figure 16 illustrates these shapes. The matrix is the main component that protects and shapes the composite. The structural ingredients include fibers, particles, laminates, flakes, and fillers, which determine the composite's interior structure. In general, but not always, structural elements are regarded as an additional phase.

The most popular types of composites are one in which the structural components are surrounded by a matrix. However, many composites lack a matrix and are made up of one or more types of elements, combining two or more materials. For example, the names "sandwich" and "laminates" refer to multi-layer arrangements that, when joined, produce a composite material. Many woven goods do not have a matrix constituent but consist of fibers with various compositions, with or without phase bonding. The matrix is the phase that provides continuity within a composite, with the dispersed/ filled/reinforcing phases embedded inside it. Typically, the matrix is a more ductile and less rigid phase. The matrix serves to hold the dispersed phases in place and share the load with the fillers. Matrices can be made of polymers, metals, or ceramics. The matrix functions are to transmit forces between the fillers, maintain their orientation, protect the fillers from environmental damage (such as heat, humidity, microorganisms, etc.), and prevent crack propagation between fillers. To effectively perform these functions, the desired material must have good ductility, toughness, high shear strength between laminates, stable temperature behavior, and high resistance to environmental humidity. Figures 17a - 17d are examples of polymer-based composite applications.

There are 5 classifications of polymer composites based on the form of the structural constituent (Callister, 2018):

1. Fiber composites consist of fibers with or without a matrix.
2. Flake composites consist of flat flakes with or without a matrix.
3. Particulate composites consist of particles with or without a matrix.
4. Filled composites consist of a continuous skeletal matrix filled with other material.
5. Laminar composites consist of constituent layers.

Figure 17. (a) Vent on automotive bumper; (b) Panel on furniture; (c) Mug made from natural fiber-filled recycle plastic; (d) Cutlery made from natural flake- filled recycle plastic.

The mixing of different materials within composite materials (in the form of a matrix and dispersed phases) always results in adjacent regions. Simply put, this refers to an interface or, in other words, a surface that forms a boundary between the constituents. In some cases, these adjacent regions are also considered an additional phase. Research on composites has primarily focused on natural fillers rather than non-natural fillers. However, non-natural fillers tend to cause abrasion on processing equipment and increase the density of a composite. Additionally, environmental issues and high production costs associated with non-natural fillers have led researchers to shift their focus to natural fillers. Natural fillers provide several advantages, including low cost, high modulus and strength, low density, ease of treatment/ modification, non-abrasion, and widespread availability.

Fillers, also known as dispersed phases, are often fibers, particles, or flakes that are added to the matrix phase to increase the mechanical and physical properties of composites, such as strength, stiffness, and toughness. Reinforcing or scattered phases can attain features such as (Schwartz, 1984):

a. Maximum improvement in physical properties.
b. Low moisture absorption.
c. Good wetting properties.
d. Low cost and easy availability.
e. Good fire resistance.
f. Good chemical resistance.
g. Low solubility in water and solvents.
h. Good heat resistance.
i. Available in various forms.

The use of natural fillers as reinforcements in composite materials provides various advantages over mineral fillers, including strength, light weight, environmental friendliness, economic feasibility, and renewability. However, natural fillers also have weaknesses and limitations, such as biodegradability due to moisture, weak surface adhesion to hydrophobic polymers, irregular filler size, unsuitability for high-temperature applications, and susceptibility to insect and fungal attacks. With the current technological advancements demanding the production of environmentally friendly and more economical products, industries are increasingly utilizing renewable natural resources. Fillers containing cellulose have received a lot of attention in the composites industry because of their potential to reinforce low-melting-point thermoplastic polymers. In addition to cellulose, another filler material that can be utilized is cellulose acetate, which is generated from the acetylation of cellulose taken from oil palm empty fruit bunches.

Conclusion

The utilization of Empty Fruit Bunches (EFB) from oil palm marks a significant step forward in sustainable waste management within the palm oil industry. Rich in cellulose (36–42%), hemicellulose (20–40%), and lignin (14–30%), EFB is an excellent raw material for various applications. These components enable EFB to be effectively utilized in the production of renewable energy, biodegradable products, and composite materials. The growing demand for natural fillers, such as cellulose from EFB, is driven by their environmental benefits, cost-effectiveness, and the potential to replace synthetic materials that are more expensive and environmentally harmful.

EFB has great potential as a sustainable source of biofuels, organic fertilizers, biodegradable pots, and high-value items including cellulose acetate and microcrystalline cellulose. EFB can also be turned into activated carbon, which is an effective adsorbent in wastewater treatment, a catalyst for biodiesel generation, and a feedstock for biochar. Despite problems like biodegradability, moisture sensitivity, and surface adherence, continuous research and technical developments are expected to improve the use of EFB.

Integrating EFB-based goods into industrial applications represents a promising step toward more sustainable and ecologically friendly operations. Among the several prospective applications, bioenergy production - particularly as a raw material for biofuels - has the most promise. This strategy not only minimizes waste in the palm oil business, but it also promotes the creation of a circular economy. By properly converting EFB into valuable goods, the palm oil business may reduce its environmental impact, generate economic benefits, and contribute to renewable energy initiatives.

Chapter 4

The Potential of Mesocarp Fiber

Abstract

This chapter explores the potential of oil palm mesocarp fiber (OPMF) as a high-value biomass resource. Derived as a by-product of palm oil processing, OPMF contains significant amounts of cellulose, hemicellulose, and lignin, making it suitable for applications such as bio-composites, bio-oil production, and renewable energy. Various pretreatment and drying methods, including superheated steam and microwave-assisted drying, are discussed in terms of their impact on the fiber's physicochemical and morphological properties. Additionally, studies on pyrolysis and co-pyrolysis demonstrate the potential of OPMF in producing bio-oil, biochar, and other renewable products. These findings highlight OPMF as an underutilized biomass that can support green, circular, and sustainable economic models.

1. Introduction

Oil palm fiber constitutes one of the principal residues produced during palm oil processing. As a secondary product of the palm oil industry (Wirman and Apriza, 2016), it is predominantly utilized as a fuel source for boilers. This material is categorized as lignocellulosic biomass, mainly comprising cellulose (59.6%), lignin (28.5%), crude protein (3.6%), fat (1.9%), ash (5.6%), and a minor fraction of impurities (8%). The fiber originates from the mesocarp portion of the oil palm fruit following mechanical extraction using a screw press, which represents a key operation within palm oil milling. Owing to its derivation from the mesocarp, oil palm fiber possesses a relatively short length that reflects the structural characteristics of the fruit tissue (Wahyono et al., 2008).

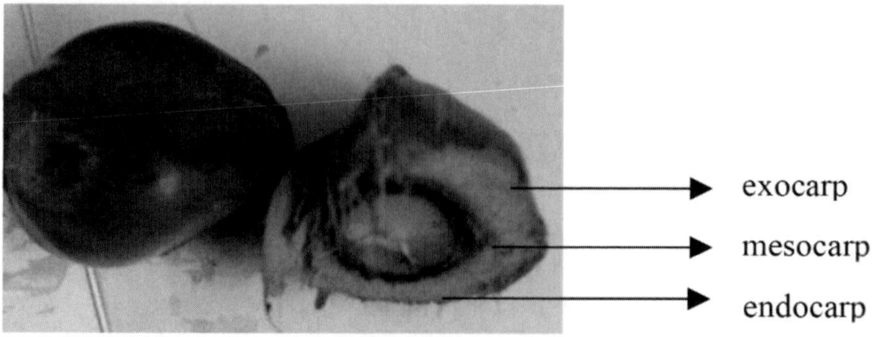

Figure 18. Mesocarp fiber.

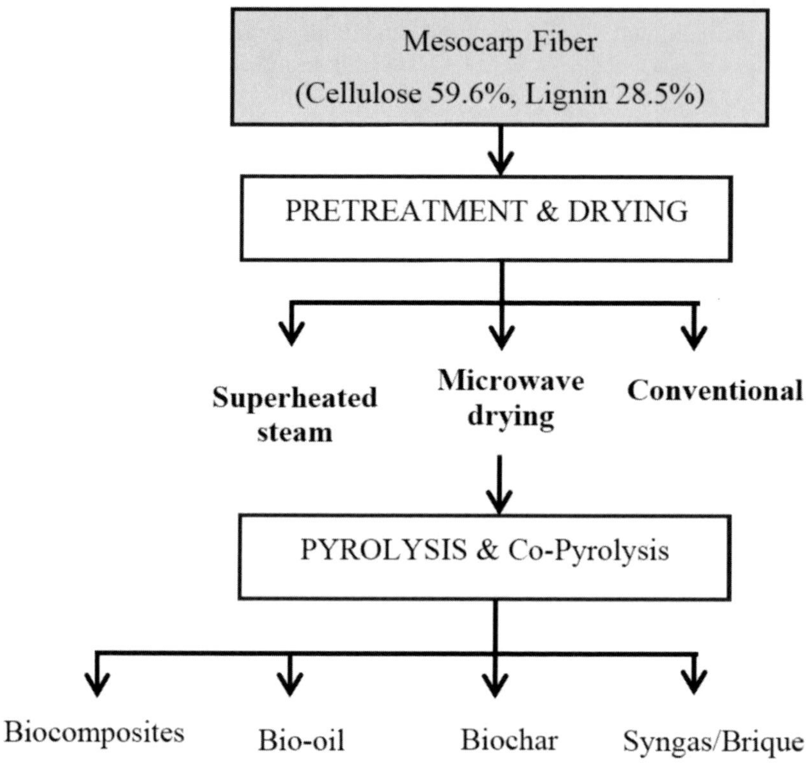

Figure 19. Mesocarp conversion process flowchart.

Figure 18 illustrates the mesocarp fiber, a key component derived from the pulp of the oil palm fruit. This fiber plays an important role in the palm oil industry, primarily as a byproduct of palm oil extraction. It is rich in cellulose and lignin, making it a valuable material for various applications. The image provides a clear view of the mesocarp fiber structure and its potential uses in promoting sustainable practices within the industry.

Oil palm fiber can also be classified as a natural fiber, making it a potential substitute for synthetic fibers. The advantage of natural fibers over synthetic ones lies in their eco-friendliness, as they are biodegradable, whereas synthetic fibers are more difficult to decompose. Natural fibers also possess the unique characteristic of being renewable (Sabarisman, 2015). However, natural fibers have certain drawbacks, such as non-uniform fiber dimensions and the influence of fiber age on their strength. The development of natural fibers as reinforcement materials for composites is highly promising, considering the abundant availability of natural fiber resources in Indonesia (Purboputro, 2017).

Figure 19 shows the mesocarp conversion process, starting with mesocarp fiber and passing through pretreatment and drying stages, which include superheated steam, microwave, or conventional drying. The processed fiber then undergoes pyrolysis and co-pyrolysis, producing biocomposites, bio-oil, biochar, and syngas/briquettes.

1.1. The Drying of Oil Palm Mesocarp Fiber (OPMF)

Oil palm mesocarp fiber (OPMF) is a lignocellulosic by-product produced as a result of the mechanical extraction process in palm oil mills. During oil extraction, the mesocarp portion of the fruit is subjected to pressing, leaving behind fibrous residues known as OPMF. In most industrial practices, this material is either left to accumulate as waste within the processing facility or incinerated to serve as a low-grade fuel source for boilers. Despite being largely underutilized, OPMF possesses notable mechanical characteristics, including a tensile strength of approximately 80 MPa, a Young's modulus of 500 MPa, and an elongation at break of around 17%. These characteristics indicate the strength and elasticity of the fiber, signifying its potential for broader applications beyond traditional energy use. Specifically, the combination of moderate strength and flexibility renders OPMF suitable as a reinforcing agent or filler in the development of natural-fiber-based biocomposites. The incorporation of OPMF in composite materials not only enhances the mechanical performance of the final product but also supports efforts toward sustainable material innovation. Moreover, the valorization of this agricultural residue contributes to the circular economy concept by transforming an abundant biomass waste into a high-value engineering material. Consequently, the utilization of OPMF in composite technology is anticipated to significantly increase the added value of palm oil biomass while reducing environmental burdens associated with waste accumulation and open burning (Then et al., 2013).

Nordin et al., (2013) conducted a study in which oil palm mesocarp fiber (OPMF) was subjected to a drying process using superheated steam technology. The results demonstrated significant improvements in the chemical composition, thermal characteristics, and morphological structure of the fiber after treatment. The superheated steam process selectively reduced the hemicellulose fraction within the fiber matrix, indicating its efficiency in modifying the surface chemistry toward a more hydrophobic nature. This transformation is particularly beneficial for enhancing the compatibility of lignocellulosic fibers in biocomposite formulations. In addition to chemical modification, the treated OPMF exhibited a notable enhancement in thermal resistance, as shown by an increase in the onset degradation temperature from approximately 258°C to 285°C. This improvement suggests that the superheated steam treatment contributed to the stabilization of the fiber's structural components. Moreover, the process effectively eliminated silica deposits from the fiber surface, leading to a rougher morphology that can promote better interfacial bonding with polymer matrices during composite fabrication. Overall, these findings confirm that superheated steam drying serves as a promising and environmentally friendly pretreatment method for producing OPMF with properties suitable for incorporation into polymer-based composite materials.

Another method that has been applied for drying OPMF is microwave drying. Microwave-assisted drying has been widely utilized because it offers several advantages, such as volumetric heating, shorter processing time, energy efficiency, and the ability to be carried out without direct contact with the material being dried. During the microwave irradiation process, energy is transferred into the material through interactions between the microwaves and the polar molecules within the material (Chang et al., 2015).

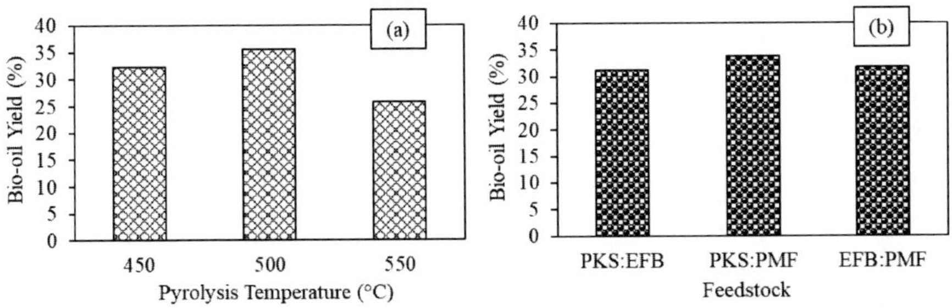

Figure 20. Bio-oil yield: (a) from pyrolysis of PMF at various temperature; (b) from co-pyrolysis of oil palm biomass (Redrawn from Kongto et al., 2025).

Chang et al., (2017) applied microwave drying to OPMF using microwave powers of 100 W and 200 W. The results showed that microwave drying reduced the mass of OPMF by 6.0% at 100 W and 18.6% at 200 W. This weight reduction was attributed to the decrease in moisture content in OPMF, which subsequently caused shrinkage in its size. However, the microscopic observation revealed that the oil glands within the oil palm mesocarp fiber (OPMF) remained largely unaffected by the microwave drying process. Analytical results indicated that microwave treatment caused a slight expansion in the fiber's cellular structure. Specifically, the average cell width increased from 0.128 mm in untreated samples to 0.143 mm and 0.1698 mm following exposure at 100 W and 200 W, respectively. This enlargement of the cell dimensions was attributed to oil redistribution and deposition occurring during microwave heating. Overall, these outcomes suggest that microwave drying induces only minor alterations in the microstructural characteristics of OPMF, thereby preserving most of its original morphological integrity.

1.2. The Pyrolysis of Oil Palm Mesocarp Fiber (OPMF)

Yeong et al., (2025) recently investigated the production yield and physicochemical characteristics of bio-oil and biochar generated through fast pyrolysis and co-pyrolysis of palm mesocarp fiber (PMF), palm kernel shell (PKS), and empty fruit bunch (EFB). The experiments were carried out using a twin-screw reactor equipped with an omega-shaped vessel, designed to ensure uniform heating and efficient biomass conversion. The co-pyrolysis configuration was highlighted as a promising approach for enhancing the valorization of oil palm residues, supporting the transition toward a green and circular bioeconomy by maximizing resource efficiency while minimizing waste generation. Both pyrolysis and co-pyrolysis processes were continuously operated under an inert atmosphere at temperatures ranging from 450 to 550°C for a duration of 60 minutes. The results indicated that the highest yield of bio-oil from the single pyrolysis of PMF was 35.61%, achieved at an optimal reaction temperature of 500°C. Under co-pyrolysis conditions, blending PMF with PKS resulted in a bio-oil yield of 33.70%, whereas the PMF and EFB mixture produced 31.68% (Figure 20).

The calorific values of the respective bio-oils obtained from PMF pyrolysis, PMF–PKS co-pyrolysis, and PMF–EFB co-pyrolysis were 8.90 MJ/kg, 18.58 MJ/kg, and 2.01 MJ/kg, respectively. Gas chromatography–mass spectrometry (GC–MS) analysis further revealed distinct compositional differences between the single and co-pyrolysis products. Acetic acid (16.23%) was identified as the predominant compound in bio-oil derived from PMF, while phenolic compounds became dominant in the mixed feedstock systems 16.17% for PMF–PKS and 13.43% for PMF–EFB co-pyrolysis. These findings emphasize the influence of feedstock interactions during co-pyrolysis on the distribution of chemical components and the quality of bio-oil produced.

Conclusion

Oil palm mesocarp fiber (OPMF) has emerged as a renewable biomass with significant potential for various industrial applications. Through appropriate pretreatment and drying optimization, its structural integrity and thermal resilience can be substantially enhanced, thereby increasing its effectiveness as a reinforcing agent in composite fabrication and as a feedstock for bioenergy conversion. Findings from pyrolysis investigations further demonstrate that OPMF can generate high-value products such as bio-oil and biochar. In a broader context, the valorization of OPMF not only mitigates waste accumulation but also elevates the economic value of palm-based biomass, supporting the development of a more sustainable and resource-efficient palm oil industry.

Chapter 5

The Potential of Palm Kernel Shell

Abstract

This chapter focuses on palm kernel shell (PKS), another abundant by-product of the palm oil industry, highlighting its composition, physicochemical properties, and potential applications. Rich in carbon, silica, and minerals, PKS has been explored as a raw material for carbon black, biochar, heterogeneous catalysts, and adsorbents. Innovative methods such as pyrolysis, gasification, and hydrothermal processes have been employed to transform PKS into functional materials for biodiesel production, wastewater treatment, energy storage, and environmental remediation. Case studies are presented on the synthesis of PKS-derived catalysts, including CaO and silica-supported composites, as well as activated carbon for gas and liquid adsorption.

1. Introduction

Palm kernel shell (PKS) is a lignocellulosic residue produced during the separation of oil palm kernels in the palm oil milling process. This material constitutes the hard outer layer of the kernel that remains after the extraction of palm oil from the mesocarp and subsequent recovery of the kernel (Figures 21 and 22). Conventionally, PKS is frequently regarded as an agricultural waste and is either discarded or subjected to open burning for disposal. In recent years, increasing scientific attention has been directed toward the valorization of PKS due to its abundance and favorable physicochemical properties. Research efforts have explored its application in diverse fields, including wastewater detoxification, use as an abrasive material in automotive manufacturing, reinforcement in polymeric and metallic composites, bioenergy generation, additive for drilling fluids, livestock feed supplement, and construction aggregates. Despite these promising applications, in rural regions, economic priorities often remain concentrated on the production of crude palm oil and palm kernel oil obtained from the mesocarp and kernel, respectively. Consequently, by-products such as PKS, palm kernel cake, palm oil mill effluent, and fibrous residues from fruit processing are still frequently disposed of without adequate management. Numerous studies have further demonstrated the versatility of PKS as a functional raw material in multiple technological domains. It has been employed as a bio-adsorbent for pollutant removal, a feedstock for biofuel synthesis, a reinforcing phase in composite materials, an abrasive medium in machining, a structural aggregate in construction, a precursor for ceramics, and a component in oil purification systems. Additionally, PKS and its derivatives have shown potential in biotechnological applications, functioning as antibacterial, antifungal, and antioxidant agents. The subsequent sections of this work will provide a more detailed discussion on the various fields in which PKS, and its modified forms have been effectively utilized.

PKS is often left as waste or openly burned as a disposal method. In rural areas, economic focus is placed more on crude palm oil and palm kernel oil extracted from the mesocarp and kernel, respectively, while waste components such as PKS, palm kernel cake, palm oil mill waste, and various fibrous residues generated from oil palm fruit processing are often disposed of without proper management. A wide range of studies has investigated the diverse potential applications of PKS, including its use as a bio-adsorbent, biofuel precursor, reinforcing material in composite systems, abrasive medium in machining operations, construction aggregate, ceramic precursor, and as a component in oil refining processes. Moreover, PKS has also attracted attention in the field of biotechnology for its bioactive properties, demonstrating potential as an antibacterial, antifungal, and antioxidant agent.

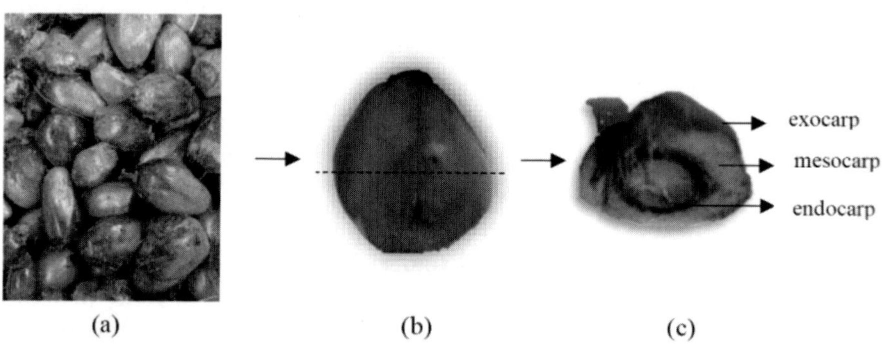

Figure 21. Photos of (a) oil palm fruit, (b) oil palm fruit, (c) cross-section of oil palm fruit.

Figure 22. Palm kernel shell processing.

1.1. The Utilization of Palm Kernel Shell as a Carbon Black

Carbon black is a finely divided black particulate material composed of over 95% elemental carbon, primarily existing as nearly spherical particles that aggregate into complex cluster structures. Recognized as one of the world's fifty most significant industrial chemicals, carbon black finds extensive use across multiple sectors, including paints and coatings, rubber reinforcement, polymer and plastic manufacturing, ink and toner production, battery electrodes, and various filler applications. Industrially, carbon black is synthesized through the controlled combustion or thermal decomposition of hydrocarbon feedstocks within a chamber supplied with a limited amount of oxygen, where natural gas commonly serves as the primary energy source to sustain the high-temperature reaction. Aromatic oils are converted into carbon black through a process of cooling and separation, resulting in fine carbon particles. Hydrocarbons decompose into carbon and gases, which are then cooled, condensed, and separated before being collected. Conventional carbon black production using fossil fuels generates significant CO_2 emissions, ranging from 2.4 to 5.7 kg of CO_2 per kg of carbon black produced. Therefore, carbon black production needs to take environmental aspects into account, and the utilization of palm oil waste, particularly palm kernel shells, offers a promising alternative feedstock.

PKS has a fairly attractive composition for utilization in carbon black production. On a dry basis, PKS contains a high amount of carbon (50.6%), along with hydrogen (6.5%), oxygen (34.5%), and nitrogen (8.4%). PKS also has relatively low moisture content (around 11–13%) and a very low ash content of only 1.4%. In addition, PKS possesses a high calorific value, with a higher heating value (HHV) of 4843.2 kcal/kg and a lower heating value (LHV) of 4499.8 kcal/kg. Several ash-forming elements in PKS, such as Si,

Ca, Fe, K, and Mg, are also present in considerable amounts. The intrinsic properties of PKS, especially its elevated carbon concentration and minimal ash content, render it an excellent precursor material for carbon black synthesis. This attribute positions PKS as a promising renewable feedstock for producing carbon black with diverse and high-value industrial applications (Aprianti et al., 2023).

Beyond its high carbon concentration, PKS possesses several advantageous properties that make it an attractive raw material for carbon black production. Firstly, PKS exhibits relatively low moisture and ash contents compared to other oil palm biomass types, which is particularly beneficial for achieving efficient carbonization and improved product purity. Secondly, its substantial carbon fraction provides a crucial foundation for generating high-quality carbon black with desirable structural and functional characteristics. Additionally, the elevated calorific value of PKS supports its dual function as both a primary feedstock and an auxiliary fuel during the production process. Moreover, PKS can undergo pyrolytic conversion to yield bio-oil, which serves as an effective alternative precursor for carbon black synthesis. Collectively, these features highlight the potential of PKS as a sustainable and renewable substitute for fossil fuel–derived materials in carbon black manufacturing, contributing to the advancement of environmentally responsible industrial practices.

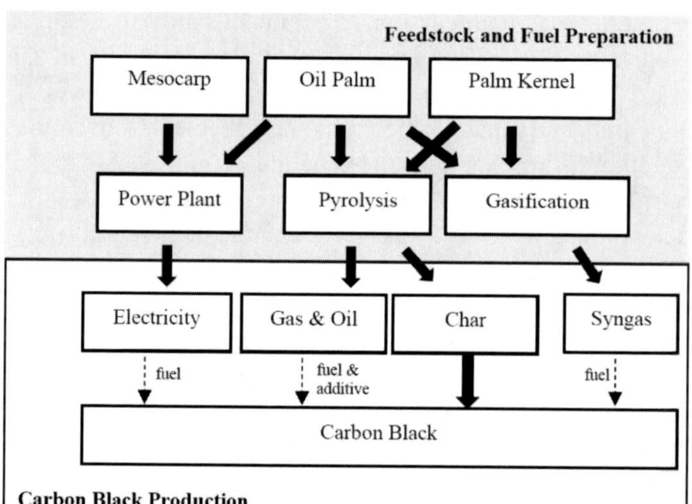

Figure 23. Schematic of carbon black production from oil palm biomass (Aprianti et al., 2023).

Figure 23 depicts the processing route for producing carbon black derived from oil palm biomass, including PKS. Prior to utilization as a precursor, the biomass feedstock must undergo conversion into oil or gaseous, which then serve as the primary raw materials for carbon black synthesis. The transformation of oil palm biomass particularly PKS can be achieved through thermochemical processes such as pyrolysis and gasification, enabling its dual function as both a carbon source and an energy provider during production. Additionally, the gaseous by-products generated from these conversion processes can be further employed as supplementary fuel, thereby enhancing the overall energy efficiency of the system.

Biomass pyrolysis is the thermal decomposition of organic materials at high temperatures in the absence of oxygen (under vacuum conditions), typically at temperatures below 500°C. The pyrolysis process is influenced by temperature and heating rate, which may vary between slow and fast processes. For example, slow pyrolysis occurs at temperatures of 300-700°C with a heating rate of 0.1-1°C/s. On the other hand, fast pyrolysis occurs at temperatures of 550-100°C with a heating rate of 100-200°C/s (Mutsengerere dkk., 2019). The breakdown of biomass generally begins with the main carbon chain, due to the lower bond dissociation energy of C–C bonds compared to C–H bonds. When heated, C–C bonds break and generate free radicals. Upon the formation of free radicals, neighboring C–H and C–C bonds become susceptible to cleavage, triggering a series of chain reactions that result in monomer degradation and volatilization. These reactive radicals may subsequently undergo secondary reactions, such as dehydrogenation or disproportionation, leading to the formation of diverse chemical compounds. At elevated temperatures exceeding 600°C, the pyrolysis of biomass predominantly generates volatile fractions, while a carbon-rich solid residue, commonly referred to as char, remains as the primary by-product (Kartik et al., 2022).

Biomass gasification refers to a thermochemical conversion process in which biomass is transformed into a combustible gas mixture under conditions of limited oxygen at elevated temperatures, typically ranging from 600 to 900°C. This process facilitates the generation of a variety of valuable products, including gaseous fuels and chemical intermediates. As a complex thermal conversion pathway, gasification integrates multiple stages, namely pyrolysis, partial oxidation, and reduction to convert solid biomass into a high-energy gas commonly known as synthesis gas (syngas). Through a combination of heterogeneous and homogeneous reactions, this syngas is predominantly composed of hydrogen (H_2) and carbon monoxide (CO) (Basu, 2018).

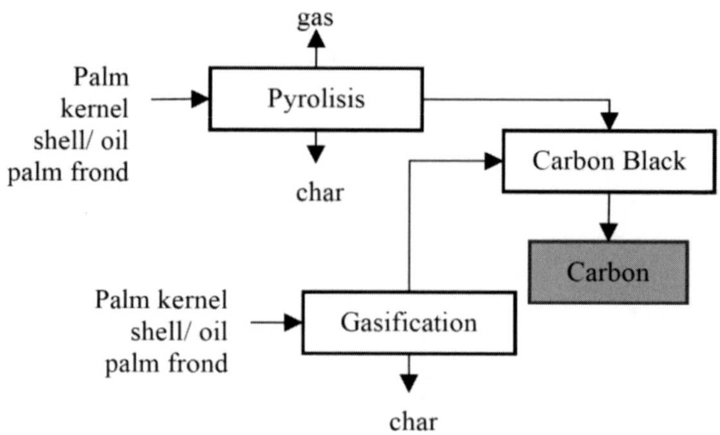

Figure 24. Schematic diagram of carbon black production (Aprianti et al., 2023).

In addition, there are two modification methods that can be applied to convert PKS into carbon black. The first is the furnace method, in which PKS is converted into pyrolysis oil that is subsequently used as a feedstock for carbon black production. The obtained pyrolysis oil is sprayed and atomized into a high-temperature furnace. Under limited oxygen conditions, part of the oil is combusted, and the resulting carbon black particles are then cooled. The second method is plasma technology, which utilizes electrical energy to create a high-temperature plasma environment for carbon black production. In this method, PKS is converted into pyrolysis oil or synthesis gas (syngas), which is then introduced into a plasma reactor. Using electrical energy, a high-temperature plasma environment is generated, and the feedstock undergoes thermal decomposition to produce carbon black. Both methods require the initial conversion of PKS into pyrolysis oil or syngas before carbon black can be produced. The choice between these methods depends on various factors, such as energy efficiency, product quality, and environmental considerations.

A representative process configuration for carbon black synthesis is illustrated in Figure 24. In this scheme, two forms of oil palm biomass are subjected to pyrolysis and gasification, resulting in the formation of char, which based on several studies can be subsequently refined or modified to enhance its properties. From a technical standpoint, the proposed method for carbon black production offers operational simplicity, as it does not necessitate major modifications to the existing reactor design or process configuration. Within this system, conventional petroleum feedstock is substituted with pyrolysis oil obtained from PKS, while the fuel source is replaced by syngas produced through PKS gasification. During operation, the pyrolysis oil is atomized and introduced into a high-temperature furnace, where it interacts with combusted syngas at temperatures reaching approximately 1500°C. Following this interaction, a portion of the pyrolysis oil undergoes controlled combustion in the presence of excess air to attain and maintain the desired thermal conditions for carbon black formation. From this process, commercial carbon black with moderate dispersion is produced within this temperature range. Several studies have noted that at higher temperatures (around 1700°C), signs of graphitization begin to appear, while at lower temperatures (around 1100°C), the carbon black structure becomes less ordered.

Various studies have also been conducted using biochar derived from pyrolysis to partially replace carbon black as a rubber filler, utilizing different types of biomasses such as rice husks, coconut shells, bagasse, dried leaves, and palm kernel shells. The resulting biochar has been evaluated in comparison with several commercial grades of carbon black, including CB N330, N772, N660, and N339. Owing to its distinct physicochemical characteristics such as a favorable specific surface area, abundant surface functional groups, stable aromatic structures, and its environmental benefit in mitigating greenhouse gas emissions - biochar is regarded as a promising carbonaceous material. Biochar possessing a high proportion of carbon and minimal ash content can be further refined into carbon black, as illustrated in Figure 25. This carbon-rich solid is generated through the pyrolysis of biomass, a thermochemical process conducted under oxygen-limited or oxygen-free conditions. The compositional and structural quality of the produced biochar, including its carbon and ash fractions, is largely influenced by the feedstock type as well as the specific operational parameters of the pyrolysis process (Greenough et al., 2021).

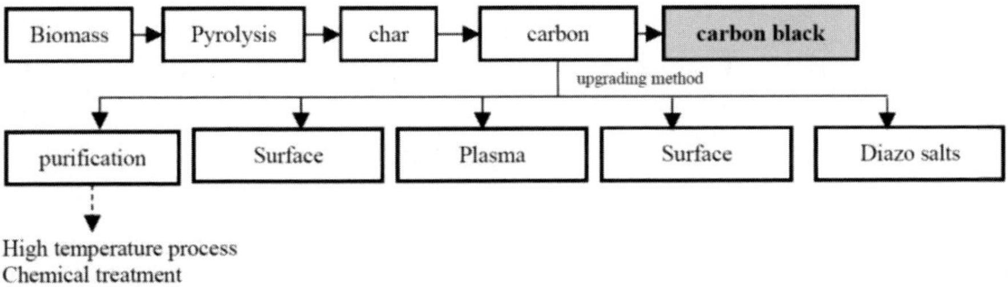

Figure 25. Schematic of carbon black production modification from biochar (Aprianti et al., 2023).

1.2. The Utilization of a Palm Kernel Shell-Based Catalyst

PKS has great potential to be reutilized as a catalyst in various reactions, one of which is the transesterification reaction in biodiesel production. The use of PKS as a catalyst is particularly attractive due to its high carbon and silica content as well as its porous physical structure, which can serve as a matrix for catalyst modification in catalytic processes, especially in the production of biodiesel from vegetable oils. Furthermore, with the increasing demand for renewable energy and efforts to reduce dependence on fossil fuels, the utilization of natural and environmentally friendly materials has become an increasingly important alternative. With proper processing, PKS can be converted into an effective catalyst to improve the efficiency of transesterification in biodiesel production.

The utilization of PKS as a solid catalyst or heterogeneous catalyst also offers many advantages in reactions. One of the primary advantages of employing heterogeneous catalysts lies in their facile separation from the reaction products, owing to the fact that the catalysts generally exist in the solid phase, whereas the reactions themselves typically proceed in the liquid phase. This enables efficient catalyst recovery, reducing costs, and minimizing product contamination. In addition, heterogeneous catalysts generally exhibit higher thermal stability, allowing them to be used under harsh reaction conditions without significant loss of catalytic activity. Another advantage is their high reusability. Solid catalysts can be reused multiple times after simple purification processes, reducing the need for new catalysts and providing both economic and environmental benefits. Heterogeneous catalysts are often regarded as more environmentally benign due to their capacity to minimize the generation of chemical waste and to facilitate the utilization of renewable or waste-derived raw materials such as palm kernel shells for the synthesis of sustainable catalytic materials. Within the scope of biodiesel production, the application of heterogeneous catalysts offers several advantages, including enhanced transesterification efficiency, reduced reliance on hazardous chemical reagents, and overall improvement in the environmental sustainability of the biodiesel manufacturing process (Narasimhan et al., 2021).

1.2.1. Palm Magnetic Biochar

Several studies have utilized the carbon and silica content in PKS, which play an important role in enhancing catalytic performance. Quah et al., (2020) processed PKS into biochar, impregnated it with Fe_3O_4 ferrite, and synthesized it using concentrated sulfuric acid through a sulfonation process. This approach aims to improve the catalytic activity of biochar, enabling its application in biodiesel production or other chemical processes. The impregnation technique involves the addition of Fe_3O_4 ferrite into the biochar to enhance its physical and chemical properties, as well as to improve adsorption capacity and thermal stability. Meanwhile, the sulfonation process introduces sulfonic groups ($-SO_3H$) onto the surface of the biochar, which can increase acidity and catalytic properties, thereby making it more efficient in various chemical reactions. After the impregnation process, cooling, neutralization, and drying are carried out, resulting in sulfonated palm magnetic biochar (PMB-SO_3H) derived from PKS.

SEM analysis revealed a rough mesoporous surface morphology with agglomerated particles of approximately 135 nm in size. After sulfonation, particle agglomeration was observed. Meanwhile, the BET surface area was lower compared to the non-sulfonated PMB catalyst. EDX analysis confirmed the successful incorporation of sulfur (S) and an increase in oxygen (O) content in the catalyst after the sulfonation process, indicating the presence of sulfonate functional groups ($–SO_3H$), as well as a reduction in iron (Fe) content. The magnetic properties of this catalyst were also analyzed using vibrating-sample magnetometer (VSM), which showed a decrease in saturation magnetization from 37.655 emu/g in the pre-sulfonated PMB catalyst to 8.458 emu/g in PMB-SO_3H, although it still exhibited superparamagnetic characteristics. Furthermore, FTIR analysis confirmed the successful incorporation of sulfonate groups and detected Fe–O bonds, indicating the transformation of ferric chloride into iron oxide. In terms of thermal stability, TGA analysis showed that the PMB-SO_3H catalyst had good stability, with no significant weight loss up to 550°C. Finally, the acid density of this catalyst was measured at 1.92 mmol/g, indicating an adequate number of active sulfonic acid sites to support catalytic transesterification reactions (Quah et al., 2020).

The obtained PMB-SO_3H was employed in optimized biodiesel synthesis through Response Surface Methodology (RSM). Four main operating parameters - reaction temperature, reaction time, methanol-to-oil molar ratio, and catalyst loading - were optimized to maximize biodiesel yield. ANOVA analysis revealed that the most influential factors on biodiesel yield were the quadratic form of catalyst loading, methanol-to-oil molar ratio, and temperature. The research revealed that a maximum biodiesel yield of 90.2% was obtained under optimized conditions specifically at a reaction temperature of 65°C, a duration of 102 minutes, a methanol-to-oil molar ratio of 13:1, and a catalyst dosage of 3.66 wt%. Furthermore, the reusability assessment of the PMB-SO_3H catalyst indicated a gradual reduction in catalytic efficiency after four successive reaction cycles, primarily attributed to the leaching of active sites and partial catalyst deactivation. Overall, the study emphasized the promising application of PKS as a precursor for the development of magnetic solid acid catalysts, which can support cost-effective and efficient biodiesel synthesis. The successful incorporation of magnetic characteristics and sulfonic functional groups into biochar-based catalysts derived from agricultural residues, coupled with process optimization using RSM, underscores the potential to achieve high biodiesel yields under relatively mild reaction conditions.

The incorporation of magnetic characteristics in the PMB-SO_3H catalyst provides notable benefits for improving the overall efficiency of biodiesel synthesis. Owing to their magnetic nature, these catalysts can be conveniently recovered from the reaction medium through the application of an external magnetic field, which greatly simplifies the separation and purification process following the reaction. Moreover, magnetic separation reduces catalyst loss during the recovery process, which in turn improves catalyst utilization efficiency in subsequent cycles. Another advantage is the ability to quickly and efficiently recover the catalyst from the biodiesel–glycerol mixture, which not only shortens recovery time but also lowers overall energy consumption, making the biodiesel production process more cost-effective and environmentally friendly (Quah et al., 2020).

1.2.2. Activated Carbon

The conversion of PKS into activated carbon presents substantial opportunities to enhance biodiesel production efficiency, particularly when employed as a catalyst during the transesterification reaction. The preparation of activated carbon from PKS typically involves high-temperature thermal activation, which develops an extensive pore network and enlarges the material's surface area, thereby improving its adsorption capacity and catalytic activity. The resulting activated carbon possesses desirable physicochemical properties such as high porosity and strong adsorption affinity toward reactant molecules that contribute significantly to optimizing the conversion of vegetable oils into biodiesel. Moreover, the use of activated carbon derived from PKS as a catalyst not only provides an environmentally friendly and sustainable alternative but also reduces reliance on homogeneous catalysts, which are less environmentally friendly.

Ausavasukhi and Sooknoi (2025) conducted a study in which PKS was employed as the precursor for developing activated carbon–based catalysts. Catalysts derived from carbonaceous materials exhibit several advantages, including high catalytic activity, cost efficiency, tunable surface functionalities, and broad applicability across various chemical processes. Generally, the preparation of solid acid carbon-based catalysts involves the carbonization of organic precursors, either through conventional or microwave-assisted pyrolysis, followed by a sulfonation step using reagents such as H_2SO_4, $(NH_4)_2Fe(SO_4)_4.6H_2O$, or 4-benzenediazonium sulfonate. In this work, hydrothermal sulfonation was carried out on a mixture of PKS and sulfuric acid at different temperatures to synthesize solid acid carbon-based catalysts. The resulting materials were subsequently utilized to catalyze the etherification of glycerol with tert-butanol. The hydrothermal sulfonation process was performed in a single step, maintaining reaction temperatures between 140 and 200°C for 24 hours. The resulting catalysts were labeled as PKS-X, where X indicates the hydrothermal sulfonation temperature used. For comparison, a catalyst prepared via carbonization followed by sulfonation (CPKS-200) was also included. Various characterization techniques were applied to analyze the catalysts, including SEM-EDX, Boehm titration, static contact angle measurement, ATR-FTIR, ion-exchange back titration, and TGA-DTG. To evaluate catalyst performance, glycerol etherification with tertiary butanol was carried out in a round-bottom flask at 30–90°C for 15–150 minutes using 0.1–0.4 g of the solid acid carbon-based catalyst. Glycerol conversion and selectivity toward various glycerol ethers were calculated to assess the effectiveness of the catalysts used.

Characterization analysis of the catalysts indicated that the hydrothermal sulfonation treatment successfully introduced several acidic functional groups onto the carbon framework, including sulfonic ($-SO_3H$), carboxylic (-COOH), and phenolic (-OH) moieties. The quantity and distribution of these acidic sites were influenced by the reaction temperature, with the catalyst synthesized at 160°C (PKS-160) exhibiting the highest total acidity and an optimal balance between hydrophilic and hydrophobic properties. Among the tested samples, PKS-160 showed superior catalytic activity, achieving the greatest glycerol conversion and selectivity toward the target glycerol derivatives, specifically di-tert-butyl ether (DTBG) and tri-tert-butyl ether (TTBG). This outstanding catalytic performance was attributed to the optimal balance between the number of acid sites and the surface hydrophobic/hydrophilic properties of the catalyst, which facilitated both reactant adsorption and product desorption.

Furthermore, this catalyst exhibited excellent recyclability, with minimal loss of activity over several reaction cycles.

The one-step hydrothermal sulfonation method for producing carbon-based solid acid catalysts offers a simpler, more cost-effective, and environmentally friendly approach compared to conventional methods that require separate carbonization and subsequent sulfonation processes. Adjusting the hydrothermal sulfonation temperature also allows control over acidity and the hydrophobic/hydrophilic balance of the catalyst, which is crucial for optimizing catalytic performance in reactions such as glycerol etherification. The PKS-160 catalyst demonstrated excellent catalytic activity, stability, and recyclability, making it a promising candidate for converting glycerol - a byproduct of biodiesel production - into value-added glycerol ethers. Furthermore, the hydrophobic/hydrophilic balance on the catalyst surface was shown to influence reaction kinetics and product selectivity, an important consideration in designing heterogeneous catalysts for biomass conversion (Ausavasukhi & Sooknoi, 2025).

In addition, Babatunde et al., (2025) conducted a study on the application of calcined palm kernel shell powder (CKSP) as a solid base catalyst for the transesterification of palm kernel oil into biodiesel. To elaborate, the research explored the potential of CKSP as a sustainable heterogeneous catalyst capable of enhancing the conversion efficiency of palm kernel oil through base-catalyzed transesterification reactions. First, palm kernel fruits were obtained, cleaned, dried, and separated to isolate the kernels from their shells. Palm kernel oil was extracted and analyzed, while the PKS was divided into two parts: one part was calcined (850°C, 3 h) to produce the CKSP catalyst, while the other was processed into activated carbon bio-adsorbent (ACB) to reduce the free fatty acid (FFA) content in the palm kernel oil. The CKSP catalyst underwent comprehensive characterization using multiple analytical techniques, including BET, FTIR, XRD, SEM, and TGA. Following characterization, biodiesel was synthesized from low free fatty acid (FFA) oil through a transesterification reaction catalyzed by CKSP. The reaction parameters, such as temperature, reaction duration, catalyst loading, and methanol-to-oil molar ratio, were optimized using the Definitive Screening Design (DSD) statistical approach. The resulting biofuel yield obtained under optimal conditions was subsequently verified against standard specifications, and the reusability performance of the CKSP catalyst was assessed through repeated reaction cycles.

The application of ACB derived from PKS demonstrated high efficiency in lowering the FFA content of palm kernel oil to 0.52% through a bioadsorption mechanism. The subsequent characterization of the CKSP catalyst revealed a BET surface area of 0.7804 m^2/g with a mesoporous structure. Elemental analysis indicated that the catalyst predominantly contained potassium, magnesium, calcium, phosphorus, and sodium, which collectively contribute to its effectiveness as a heterogeneous base catalyst in biodiesel synthesis. Optimization of biodiesel production using DSD approach predicted an optimal biofuel yield of 99.73% (wt/wt) at a reaction time of 67.574 minutes, a reaction temperature of 67.116°C, a CKSP concentration of 3.957% (wt.), and a methanol-to-oil molar ratio of 1:7.989. The properties of the resulting biofuel, such as density, viscosity, water content, acid value, iodine value, saponification value, peroxide value, cetane number, and calorific value, met the recommended ASTM and EN biofuel standards. In addition, the CKSP catalyst demonstrated good recyclability, with only a slight decrease in biofuel yield after several cycles, indicating its potential for industrial-scale application (Babatunde et al., 2025).

1.2.3. Silica

An emerging innovative application is the use of PKS as a raw material for silica-based catalysts. The hard structure of PKS and its inorganic composition, which is rich in silica, allow this material to be transformed through specific chemical processes into a heterogeneous catalyst. Silica-based catalysts derived from PKS not only provide a solution to waste management issues but also strengthen national chemical industry independence by offering catalysts that are economical, environmentally friendly, and locally accessible. Moreover, the utilization of silica-based catalysts derived from PKS holds significant potential for various industrial-scale applications, including esterification and transesterification reactions in biodiesel production (Ngaini dkk., 2020), as well as other organic transformation processes that demand high catalytic efficiency and selectivity. This advancement also aligns with the principles of the circular economy, where industrial by-products are redefined as valuable raw materials, transforming waste into a strategic asset that enhances product value and strengthens the global competitiveness of national industries.

Several researchers have developed silica-based catalysts derived from PKS. Novita et al., (2024) reported the synthesis of an eco-friendly and reusable heterogeneous catalyst obtained from PKS ash (PKSA_DHC) through a straightforward solid-state preparation method. The high silica content present in PKS ash makes it a promising and sustainable precursor material for catalyst production in biodiesel synthesis. Moreover, the catalytic performance of PKS ash can be further improved by forming heterogeneous composite catalysts, which enhances the efficiency of transesterification reactions. Moreover, other mineral contents such as K_2O in PKS ash also contribute to the improvement of its catalytic properties. This can be observed in the elemental and compound composition of PKS ash presented in Table 4.

The research stages began with PKSA obtained by burning PKS at 1000°C in a palm oil mill boiler. To prepare the PKSA_DHC catalyst, PKSA was sieved, treated with 1 N HCl to remove impurities, and dried. The PKSA_DHC catalyst was synthesized via a solid-state technique, which involved mixing, grinding, and subsequent calcination of PKSA with NaOH at varying temperatures (350°C, 400°C, 450°C, and 500°C) for a duration of two hours. The resulting catalyst was characterized through XRD, FTIR as well as SEM to determine its structural and morphological properties. For the transesterification experiments, Response Surface Methodology (RSM) based on a Central Composite Design (CCD) was utilized to optimize critical reaction parameters such as catalyst loading, methanol-to-oil molar ratio, reaction temperature, and time. The biodiesel synthesis was then conducted under optimized conditions, and the corresponding yield was evaluated. After completion, the spent PKSA_DHC catalyst was recovered, thoroughly washed with n-hexane, and dried for reuse in subsequent transesterification cycles. The biodiesel produced was analyzed for FAME composition using GC-MS, and the biodiesel characteristics were benchmarked against EN 14214 and ASTM D6751 standard specifications.

The key characteristics of the PKSA_DHC catalyst analyzed indicate that the catalyst calcined at 450°C exhibited the most optimal structural and compositional properties. XRD analysis showed that at calcination temperatures of 350°C and 400°C, the catalyst still displayed SiO_2 peaks, suggesting incomplete formation of the Na_2O/SiO_2 composite. However, at 450°C and 500°C, the SiO_2 peaks disappeared, and sharp peaks corresponding to the Na_2O/SiO_2 composite appeared, indicating optimal catalyst structure formation at 450°C. FTIR analysis revealed the presence of Si–O–Si and Si–O–Na bonds in all catalysts, with the highest intensity of Si–O–Na bonds observed in the catalyst calcined at 450°C, demonstrating the most optimal formation of these functional groups. Morphological analysis revealed that PKS ash had non-uniform particle sizes that tended to form clusters, whereas the PKSA_DHC catalyst calcined at 450°C showed more uniform particle sizes, distinct grain boundaries, and favorable pore structures that facilitate methanol access to the catalyst's active sites. Overall, these characterization techniques demonstrated that the catalyst calcined at 450°C possessed optimal pore structure and morphology, supporting transesterification reactions with higher efficiency.

The findings of the study indicated that the optimal parameters for biodiesel synthesis were achieved with a catalyst loading of 6 wt%, a methanol-to-UPCO molar ratio of 15:1, a reaction temperature of 50°C, and a reaction duration of 90 minutes. Under these optimized conditions, a high biodiesel yield of 99.01% was obtained. The resulting biodiesel fulfilled the physicochemical requirements established by the EN 14214 and ASTM D6751 standards. Additionally, the catalyst demonstrated good reusability, maintaining a biodiesel yield of over 80% after six consecutive cycles. The application of PKS ash as a catalyst precursor also enabled a 47.24% reduction in PKS ash waste, highlighting its environmental and economic advantages (Novita et al., 2024).

Manurung et al., (2023) also investigated PKS-derived silica as a matrix for nano-CaO catalysts in biodiesel production. Silica was employed as a support material for CaO catalysts because CaO suffers from poor stability during transesterification reactions, leading to the leaching of Ca^{2+} ions. The problem can be effectively resolved through the application of a wet impregnation method on the CaO catalyst support, thereby reducing the leaching of Ca^{2+} ions and enhancing the catalyst's stability throughout the reaction.

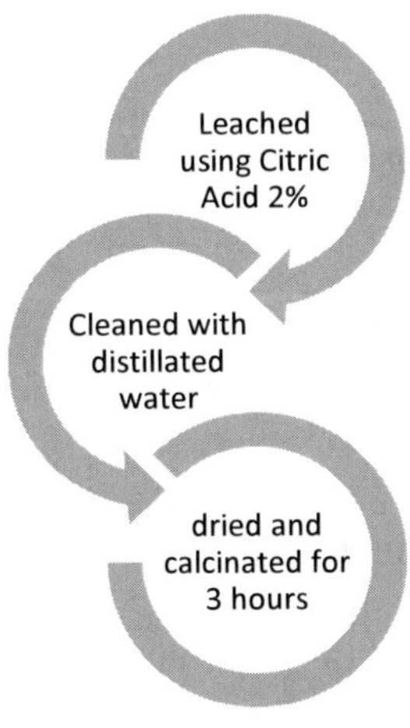

Figure 26. Processing of PKS Ash into Silica (Manurung et al., 2023).

Table 4. Elemental and compound composition of PKS ash (Novita et., 2024)

Element	Composition (%)	Compound	Composition (%)
Si	74.9	SiO2	86.15
K	10.52	K2O	5.01
P	4.07	P2O5	3.89
Ca	4.06	CaO	2.08
Fe	3.92	Fe2O3	1.94
Others	2.53	Others	0.93

Manurung et al., (2023) also investigated PKS-derived silica as a matrix for nano-CaO catalysts in biodiesel production. Silica was employed as a support material for CaO catalysts because CaO suffers from poor stability during transesterification reactions, leading to the leaching of Ca^{2+} ions. The problem can be effectively resolved through the application of a wet impregnation method on the CaO catalyst support, thereby reducing the leaching of Ca^{2+} ions and enhancing the catalyst's stability throughout the reaction.

In brief, the preparation process of the nano-CaO/SiO_2 catalyst is illustrated in Figure 26. The study began with the extraction of nano-CaO from chicken eggshells using the CHD (Calcination–Hydration–Dehydration) method, with initial calcination at 900°C to produce CaO, followed by hydration and further calcination at 870°C to obtain nano-CaO. PKS ash was treated with a 2% citric acid solution, rinsed, dried, and calcined at 700°C to extract silica. The nano-CaO was then impregnated with palm ash at various concentrations (10–50 wt%), mixed, dried, and calcined at 600–900°C (as shown in Figure 26). The resulting catalysts were characterized using PSA, FESEM, XRD, FTIR, BET-BJH, and Hammett indicator titration. Biodiesel transesterification was conducted using palm oil, methanol, and the catalyst at 60°C, with a methanol-to-oil ratio of 15:1 and stirring at 500 rpm. Reaction time and catalyst loading were varied to optimize biodiesel yield. The biodiesel obtained was analyzed for methyl ester content, density, viscosity, and other properties according to the EN14214 standard.

The physicochemical characteristics of the nano-CaO/SiO_2 catalyst have a direct impact on the resulting biodiesel yield. The nano-CaO particles exhibited an average diameter of approximately 24.98 nm, while the catalyst incorporating 20 wt% PKS ash and calcined at 800°C demonstrated superior textural properties, including the highest specific surface area (21.483 m^2/g), pore volume (0.057 cc/g), and basicity level (8 mmol/g). However, increasing the PKS content above 20 wt% reduced both alkalinity and catalytic activity, as the active sites of nano-CaO became covered by Si compounds from the PKS ash, thereby lowering the total alkalinity. The catalyst calcined at 800°C demonstrated the best catalytic performance due to its optimal surface area and alkalinity. Conversely, calcination temperatures above 800°C decreased both surface area and alkalinity. The correlation between catalyst properties and biodiesel yield indicated that the catalyst with higher alkalinity and surface area (20 wt% PKS ash, 800°C) achieved the maximum biodiesel yield of 71.77%, while higher PKS ash loading and elevated calcination temperatures reduced surface area, alkalinity, and overall biodiesel yield.

Table 5. Performance comparison of PKS-derived catalyst

Catalyst Type	Preparation Method	Catalyst Characteristics	Catalyst Performance
Magnetic biochar-based catalyst (Quah et al., 2020)	Pyrolysis of PKS into biochar, biochar impregnation using Fe3O4 ferrite, followed by sulfonation using H2SO4	The catalyst has a mesoporous surface with particle agglomeration The catalyst contains -SO3H groups Saturation magnetization of 8.458 emu/g Acid density of 1.92 mmol/g	Biodiesel yield of 90.2%, reusability of 4 times
Activated carbon-based catalyst (Ausavasukhi and Sooknoi, 2025)	Hydrothermal sulfonation of PKS	The catalyst contains -SO3H, -COOH, and -OH groups	
Calcined PKS-based catalyst (Babatunde dkk., 2025)	Calcination	The catalyst has a mesoporous structure with surface area of 0.7804 m2/g The catalyst contains K, Mg, Ca, P, dan Na	Biodiesel yield of 99.73%
Silica-based catalyst (Novita dkk., 2024)	Solid-state preparation	The catalyst shows the presence of Si-O-Si and Si-O-Na groups The catalyst has a non-uniform particle morphology	Biodiesel yield of 99.01%, reusaibility of 6 times
Nano-CaO/SiO2 catalyst	Nano-CaO preparation by calcination-hydration-dehydration (CHD) method Silica extraction by sol-gel method Catalyst preparation by wet impregnation approach	Surface area of 21.483 m2/g and pore volume of 0.057 cc/g Basicity of 8 mmol/g	Biodiesel yield of 74.4%.

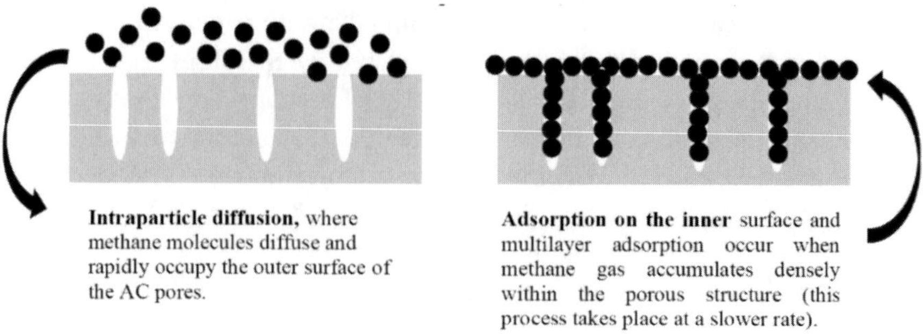

Figure 27. Adsorption mechanism of PKS-based activated carbon (Zaini et al., 2023).

This study also examined the effect of catalyst loading (2–5 wt%) and reaction time (60–180 minutes) on biodiesel yield. The highest biodiesel yield of 74.4% was obtained using 3 wt% catalyst with a reaction time of 60 minutes. The physicochemical characteristics of the resulting biodiesel, such as methyl ester composition, density, and viscosity, were evaluated and confirmed to meet the requirements of the EN14214 standard. The application of waste-derived materials, including eggshells and palm kernel shell ash, in developing an efficient heterogeneous catalyst for biodiesel production highlights a sustainable and environmentally responsible strategy. A systematic strategy in catalyst preparation, characterization, and optimization of process parameters was employed to achieve high biodiesel yield (Manurung et al., 2023).

Overall, Table 5 summarizes the four studies utilizing PKS as a precursor for heterogeneous solid catalysts. All four studies demonstrated satisfactory catalyst performance, with product yields >90%. The catalyst preparation procedure influences catalyst characteristics - for example, impregnation with a compound during preparation results in functional groups on the catalyst surface.

1.3. The Utilization of Palm Kernel Shell as an Adsorbent

Recent research has revealed that PKS possess physical and chemical characteristics with great potential to be developed into adsorbent materials. Adsorbents are materials capable of binding or adsorbing pollutants from liquids and gases, making them highly important in water and air purification processes, as well as in industrial wastewater treatment. With high porosity, large specific surface area, and a favorable composition of carbon and minerals, PKS can be processed through carbonization and activation into high-quality activated carbon. In addition to being environmentally friendly, PKS-based adsorbent products offer an economical solution while reducing pollution from palm biomass waste, which is often not optimally managed. The development of adsorbent technology from PKS not only broadens the alternatives for activated carbon feedstock at the national scale but also strengthens the competitiveness of Indonesia's palm oil industry at the global level. Converting waste into high-value adsorbents is multifunctional, ranging from filtering heavy metals, pesticides, and other pollutants in water and air, to applications in health, agriculture, and the food and beverage industries. This transformation signifies the adoption of green and innovative technologies, while also reinforcing the role of palm oil as a strategic resource that supports Indonesia's ecological and economic sustainability.

Several studies have been conducted on the development of adsorbents from PKS. Zaini et al., (2023) investigated the synthesis of activated carbon from PKS using various activating agents, namely CO_2, steam, and a combination of CO_2 and steam. PKS was first impregnated with a zinc chloride ($ZnCl_2$) solution and then carbonized under a nitrogen (N_2) flow at 850°C. Subsequently, the carbonized samples were treated with CO_2, steam, or a combination of CO_2 for one hour followed by steam for one hour for the activation process. The produced activated carbon was characterized to determine surface area, pore volume, and pore size distribution using the BET method and SEM. In addition, methane adsorption was measured using the volumetric method at room temperature, and the amount of methane uptake was calculated using the ideal gas equation.

The methane adsorption efficiency of activated carbon (AC) is largely governed by its pore structure characteristics, which are influenced by factors such as carbon burn-off, yield, surface area, pore volume, and pore diameter. The selection of activating agents plays a crucial role in determining the extent of carbon burn-off during activation, where the combined use of CO_2 and steam achieved the highest burn-off rate of 78.57%, effectively enhancing pore structure development. Activated carbon synthesized using the CO_2–steam mixture (AC-Mix) exhibited the greatest surface area of 869.82 m^2/g and a total pore volume of 0.47 cm^3/g, leading to the highest methane adsorption capacity of 4.500 mol/kg. The adsorption mechanism of PKS-derived activated carbon is illustrated in Figure 27. The adsorption isotherms of all AC samples were best represented by the Freundlich model, signifying reversible, non-ideal, and multilayer adsorption characteristics. Kinetic analysis using the pseudo-first-order model revealed that adsorption is primarily governed by physical interactions between the adsorbent and adsorbate. Furthermore, intraparticle diffusion modeling indicated that methane adsorption occurs through diffusion within the internal pores as well as across the external surface due to variations in adsorption rates (Zaini et al., 2023).

Muhammad et al., (2022) investigated the application of activated carbon derived from PKS as an adsorbent for the advanced treatment of biologically processed of POME. The use of PKS-based activated carbon offers a cost-efficient alternative due to the abundant availability of PKS as an agricultural byproduct. In their study, the activated carbon (PKSAC) was modified through iron oxide impregnation to develop a magnetic adsorbent (PKSAC-Fe), allowing for convenient separation from the treated effluent. Comprehensive characterization using BET, SEM/EDX, FTIR, and VSM techniques revealed that the adsorbent possessed a high specific surface area, strong magnetic behavior, and multiple functional surface groups, confirming its suitability for wastewater treatment applications.

Batch adsorption tests were conducted to determine the optimal operational conditions, including initial pH, adsorbent dosage, and contact time, utilizing the Box–Behnken experimental design approach. The optimized parameters, initial pH of 3.5, adsorbent dosage of 9.9 g/L, and contact duration of 3.4 hours, enabled PKSAC-Fe to achieve remarkable removal efficiencies of 99.7% for color and 85.0% for COD from biologically treated POME. Moreover, the adsorbent exhibited strong regeneration performance, maintaining high removal efficiency after four consecutive adsorption–desorption cycles with only minimal decline. The application of PKS-derived adsorbents in POME treatment supports the concept of cleaner production and circular economy by promoting waste valorization, resource recovery, and environmental sustainability (Muhammad et al., 2022).

Conclusion

Palm kernel shell (PKS) represents a highly valuable and multifunctional form of lignocellulosic biomass with considerable potential for diverse industrial applications. Owing to its rich carbon composition and substantial silica content, PKS serves as an excellent precursor for the synthesis of advanced carbon-based materials such as carbon black, activated carbon, and heterogeneous catalysts. Numerous studies have demonstrated that materials derived from PKS exhibit superior physicochemical properties, such as high surface area, good thermal stability, and abundant functional groups, making them efficient, recyclable, and environmentally benign alternatives to conventional synthetic materials. The utilization of PKS-derived catalysts and adsorbents not only supports the transition toward greener industrial processes but also enhances the overall sustainability of the palm oil value chain. Through systematic valorization of PKS, the palm oil sector can significantly reduce biomass waste, produce high-value functional materials, and strengthen its alignment with the principles of circular economy and sustainable resource management.

Chapter 6

The Potential of Oil Palm Fronds and Stems

Abstract

Oil palm fronds (OPF) and trunks (OPT) are biomass leftovers produced by the palm oil plantation sector with numerous possible applications. This chapter presents a comprehensive assessment of recent research on the valuation of OPF and OPT using a variety of methodologies. OPF has been utilized for the synthesis of nanocatalysts for biodiesel production, as well as processed through torrefaction, bioethanol production, and supercapacitor development for electrical applications. Moreover, both OPF and OPT have been extensively studied for drying processes aimed at producing high-quality biomaterials, with evaluations of different drying equipment and modeling of drying kinetics and activation energy. Pyrolysis approaches have also been applied to convert OPF and OPT into bio-oil and biochar, complemented by kinetic modeling using thermogravimetric data. Overall, the valorization of OPF and OPT represents a strategic pathway to support renewable energy production, the development of functional materials, and the enhancement of the value of palm oil industry residues.

1. Introduction

Oil palm fronds (OPF) are one of the most common solid biomass leftovers found near oil palm fields and processing mills. Traditionally considered low value or even an environmental burden, OPF is currently recognized as a potential resource capable of being transformed into value-added products. Utilizing OPF as a biomass feedstock provides opportunities to reduce environmental waste while supporting various industrial sectors, considering that biomass can absorb CO_2 emissions and can be transformed into bioenergy (Kalak, 2023). This shift in mindset from trash to resource is consistent with the ideals of a circular economy and sustainable development. Bahan biomassa pertanian memiliki keunggulan dalam menyerap peningkatan kadar CO2 di atmosfer melalui penangkapan CO2 biologis. Limbah biomassa sebagai sumber energi terbarukan dapat diubah menjadi bioenergi menggunakan berbagai teknologi [6]. Strategi Uni Eropa tentang biofuel menyatakan bahwa teknologi baru untuk memperoleh dan menggunakan bahan bakar terbarukan perlu dikembangkan, terutama melalui pengelolaan produk sampingan pertanian dan limbah lainnya, termasuk limbah industri yang dapat terurai secara hayati.

OPF can be utilized as a bioenergy feedstock for electricity generation. With a calorific value of approximately 17 MJ/kg, OPF has the potential to produce up to 4.72 kWh/kg of electricity, making it an environmentally friendly and renewable biomass fuel. Conversion of OPF into biochar also offers significant benefits as a soil amendment. The resulting biochar contains high fixed carbon (25.7%) and low ash content (6.32%), enhancing soil fertility and stabilizing soil carbon through the conversion of organic carbon. Furthermore, OPF possesses sufficient nutritional value to be used as ruminant feed. With fiber content reaching 70%, soluble carbohydrates at 22%, crude protein at 4.2%, and urea levels between 1–2%, oil palm fronds can support digestion and animal health, particularly for cattle and goats. Thus, OPF not only serves as a by-product of the plantation sector but also represents a multifunctional resource contributing to various environmentally friendly and sustainable industrial applications (Zakaria et al., 2024).

1.1. The Production of an Oil Palm Fronds-Based Catalyst

Several studies have been conducted on the utilization of OPF. Younis Al-Ani et al., (2025) studied the development of a novel green nanocatalyst for biodiesel production by impregnating activated carbon derived from OPF with zirconium oxide nanoparticles. The PFAC-ZrO_2 catalyst was created utilizing a unique impregnation process, as shown in Figure 28. Acetic acid was used to functionalize the PFAC surface, with oleic acid acting as a surfactant to disseminate the ZrO_2 nanoparticles. The resultant catalyst was then calcined at 500°C. The obtained catalyst had a large surface area of 231.22 m^2/g and an average pore size of 4.09 nm, indicating its potential for efficient catalytic activity. The Taguchi design and ANOVA were used to optimize the transesterification process for biodiesel generation, with reaction time being the most important element. Under ideal conditions, waste olive oil as a feedstock produced a maximum biodiesel yield of 97.36%.

This breakthrough not only addresses agricultural waste management challenges, but it also provides a cost-effective and ecologically beneficial technique of producing biodiesel. The PFAC-ZrO_2 catalyst showed outstanding reusability, keeping 84.8% of its activity after five cycles, thus increasing its economic viability. Furthermore, the study emphasizes the possibility for using local agro-waste in sustainable energy applications, which is consistent with the circular economy and green chemistry concepts. The focus on reaction parameter optimization and catalyst performance analysis provides valuable insights for the potential industrial implementation of PFAC–ZrO_2 derived from oil palm fronds (Younis Al-Ani et al., 2026).

1.2. Energy Generation from Oil Palm Fronds

Milicevic et al., (2025) studied the influence of torrefaction, a moderate pyrolysis process, on the anaerobic co-digestion of OPF biomass and palm oil mill effluent (POME). OPF was torrefied at 240–350°C for 30 minutes, and the impact of using torrefied OPF as a co-substrate in anaerobic co-digestion with POME was evaluated. The study analyzed physicochemical changes in the biomass due to torrefaction, as well as methane production, and the removal efficiency of various parameters during anaerobic co-digestion.

The findings revealed that increasing the torrefaction temperature lowered the bulk yield of torrefied biomass from 80 to 50 wt%. FTIR measurement revealed a decrease in the strength of peaks representing hydroxyl, carbonyl, and ether functional groups, indicating chemical composition alterations in the biomass. Co-digestion of POME with torrefied OPF at 280°C resulted in the maximum methane output of 1.36 L CH4/g VS, reflecting a 209% increase over raw OPF and POME. Furthermore, anaerobic co-digestion demonstrated remarkable removal efficiency for a variety of parameters, ranging from 77% to 99%. Torrefaction was included as a pretreatment step prior to anaerobic co-digestion, resulting in considerable improvements in methane output and substrate biodegradability. This approach provides a viable alternative for valorizing palm biomass and POME, therefore contributing to sustainable waste management in the palm oil sector (Milicevic et al., 2025).

Abdullah et al., (2021) used OPF as a renewable feedstock for bioethanol synthesis. As a plentiful byproduct of the palm oil industry with a high fermentable sugar content, OPF is a prospective feedstock for biofuel generation. However, one significant difficulty is the quick loss of fermentable carbohydrates during storage. This study looked into the influence of OPF concentration and storage at moderate temperatures on glucose content in OPF solutions. OPF was concentrated by evaporation of 30-70% of its water content, then kept at 30-60°C for up to 20 days. The results showed that glucose content remained stable when held at 50°C and 60°C, independent of concentration, but it decreased fast at 30°C and 40°C.

The study also examined bacterial growth and pH changes in stored OPF solutions. At 50°C, bacterial counts gradually decreased over time, with only thermophilic bacteria surviving after six days. At 60°C, all bacteria were eliminated within two days. The pH of the OPF solution at 60°C dropped from 4.8 to around 3.9, most likely due to the production of organic acids and Maillard reaction products. Storing a 50% concentrated OPF solution at 50°C for up to 20 days proved to be an effective preservation method to maintain glucose content and control microbial growth, making it suitable as a feedstock for bioethanol production. This moderate-temperature storage approach is more energy-efficient compared to traditional low-temperature storage methods. These findings confirm the significant potential of OPF as a feedstock for bioethanol production (Syed Abdullah et al., 2021).

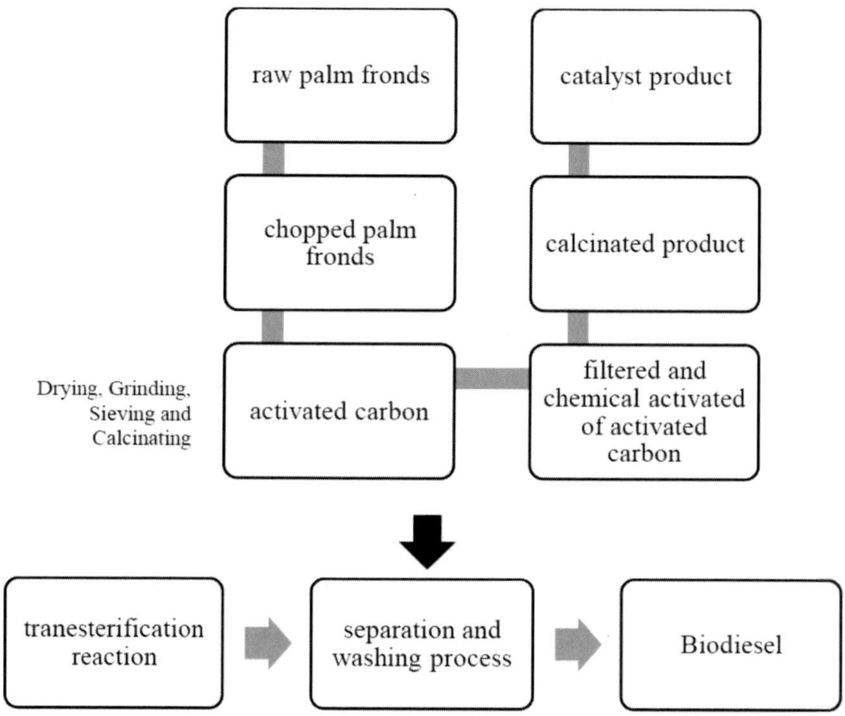

Figure 28. PFAC-ZrO2 Preparation nanocatalyst and application in biodiesel production (Younis Al-Ani et al., 2026).

OPF has also been explored in the field of electricity generation. Paramudita et al., (2025) investigated the use of OPF as activated carbon (AC) for supercapacitor applications. The researchers used a two-stage carbonization method followed by chemical activation with KOH or NaOH at different temperatures (600°C, 700°C, and 800°C). The characterization results revealed that higher activation temperatures increased the porosity and surface area of the AC. AC prepared using KOH activation at 700°C had the highest specific surface area of 2298.1 m^2/g, while AC prepared with NaOH activation at 800°C had the highest specific surface area of 2054.6 m^2/g.

Symmetric two-electrode supercapacitor devices were built with AC electrodes produced from OPF, and an H_2SO_4-PVA gel electrolyte displayed exceptional electrochemical performance. The supercapacitor using AC-NaOH-800 electrodes had the maximum specific capacitance of 490.0 F/g, energy density of 68.1 Wh/kg, and power density of 2083.3 W/kg. These results outperform many previously reported biomass-based supercapacitor systems, showing OPF's promise as a low-cost and sustainable predecessor to high-performance electrode materials.

1.3. The Drying of Oil Palm Fronds

Mohideen et al., (2011) proposed a swirling fluidization technique for drying OPF. This method is capable of efficiently drying OPF with irregular shapes while maintaining relatively low energy consumption. In their research, a laboratory-scale swirling fluidized bed dryer (SFBD) was built to test the hydrodynamics and drying properties of OPF segments in hot air flow at 45°C and 60°C. The OPF was made up of leaves (57% wet basis) and stems (70% wet basis). The results showed that drying to a target moisture content of 15% (appropriate for animal feed production) took about 60 minutes for leaves and 90 minutes for stems. The hydrodynamics and drying kinetics of the leaves and stems differed significantly; leaves dried quickly, with a considerable initial moisture loss before slowing, whereas stems dried more slowly and through diffusion. These findings indicate that the drying of leaves is reaction-rate controlled, while stem drying is diffusion-controlled. Therefore, leaves and stems should be dried separately. This type of dryer is suitable for both continuous and batch operations.

Puspasari et al., (2012) studied the drying kinetics of OPF fibers using a fluidized bed dryer assisted by mechanical agitation. Fibrous materials like OPF are difficult to dry in a fluidized bed due to fiber entanglement, which can create channeling within the bed. Mechanical agitation was shown to effectively eliminate channeling, facilitating fluidization. Using the Taguchi approach, we evaluated the impacts of inlet air temperature, superficial air velocity, bed load, and agitation speed. The drying kinetic model's appropriateness was evaluated using statistical criteria, such as modified R^2 and Root Mean Square Error (RMSE). The results showed that greater inlet air temperatures and velocities, along with a lower bed load, reduced drying time. Within the tested parameter range, drying OPF particles to 0.11 g water/g (dry basis) took 14-45 minutes. The drying rate rose with increasing air temperature and velocity and decreased bed load, but the influence of agitation speed was insignificant, indicating that good particle mixing had already occurred. A new composite model was proposed to predict OPF drying curves using constants $n < 1$ and $m>1$.

As additional research, Puspasari et al., (2014) investigated the drying properties of empty fruit bunch (EFB) fibers using a fluidized bed drier with mechanical agitation. The concept of characteristic drying curves was applied to describe the drying behavior of various materials, from edible to inedible. Results showed that the drying curve shapes of EFB did not significantly differ across the tested temperature range (115–170°C) and steam velocities (0.35–0.49 m/s). The drying rate curves showed an initial accelerating rate period then a falling-rate period, which is similar to EFB fiber drying using superheated steam, as reported by Hasibuan and Daud (2009).

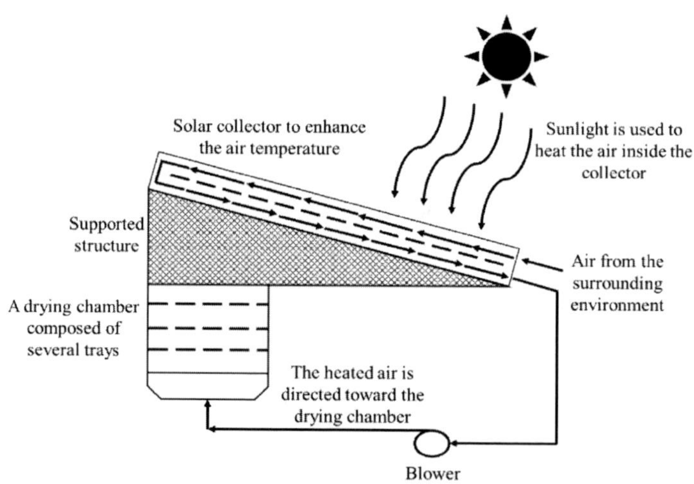

Figure 29. General schematic of indirect forced-convection solar dryers by employing solar collector.

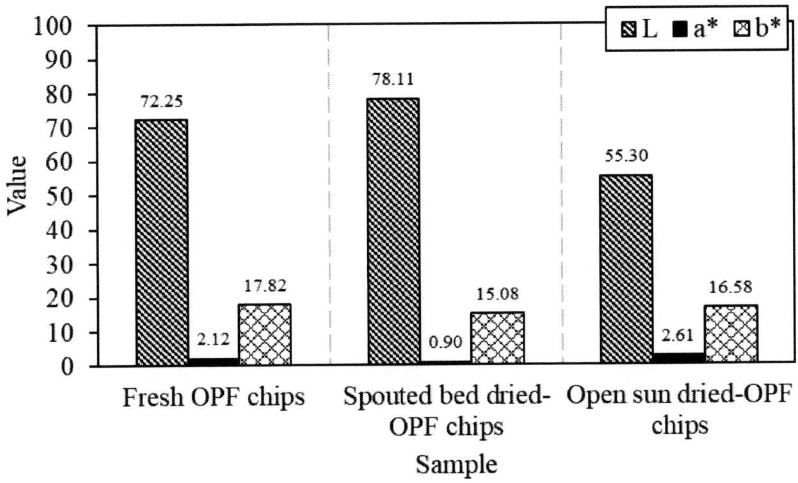

Figure 30. Color parameter of OPF chips (Tan et al., 2019).

Despite advances in dryer technology, indirect forced-convection solar dryers are still used for palm biomass, as illustrated in Figure 29. Fudholi et al., (2015) tested an indirect forced-convection sun drying system for 100 kg of OPF. The dryer reduced OPF moisture content from 60% (wb) to 10% (wb) over 22 hours (three days of drying). During drying, the average daily inlet air temperature ranged from 26–75°C, outlet air temperature from 25–65°C, internal dryer air temperature from 26–67°C, and solar radiation from 96–1042 W/m^2. The system has a specific moisture extraction rate (SMER) of 0.29 kg/kWh and an average exergy efficiency of 47%. Furthermore, the average potential improvement of the solar drying system for OPF was estimated at 172 W.

Tan et al., (2019) investigated the drying kinetics and characteristics of OPF particles using a spouted bed dryer. The study demonstrated that the drying rate was directly proportional to the inlet air temperature and air velocity, but inversely proportional to the particle bed height. Analysis of the OPF particle drying kinetics using drying curves revealed an initial drying period followed by two falling-rate periods, but no significant constant-rate period was observed. The best operating conditions were determined to be an air temperature of 90°C, an air velocity of 1.34 m/s, and a bed height of 10 cm. Furthermore, drying with the spouted bed produced OPF particles with brighter color (as shown in Figure 30), improved aroma, and smoother surfaces compared to open-sun drying.

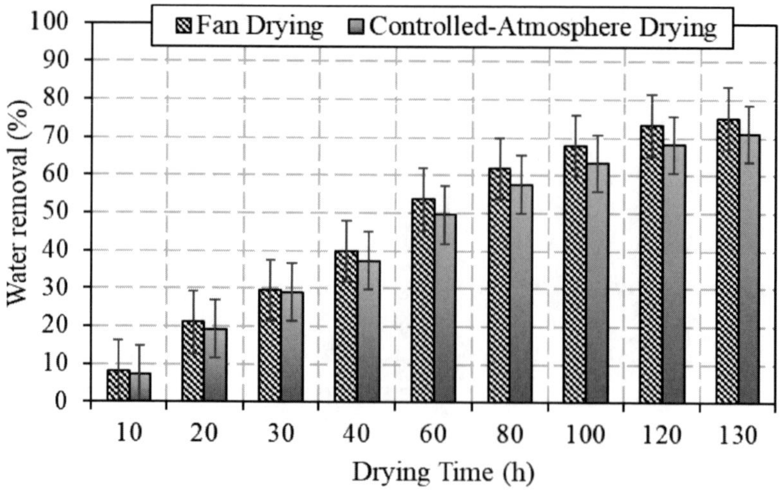

Figure 31. Water removal percentage (Redrawn from Nadhari et al., 2014).

1.4. The Drying of Oil Palm Trunks

Nadhari et al., (2014) investigated the drying kinetics of oil palm trunks (OPT) with a laboratory-scale convective dryer under two different operating conditions: open-air fan drying and controlled-atmosphere drying (25°C, 80% relative humidity). These two settings were chosen to assess the impact of keeping a constant atmosphere in terms of relative humidity and temperature on drying pace. As shown in Figure 31, several data points, particularly from the fan drying experiments, fluctuated around the trend line due to variations in temperature and relative humidity during the drying process. In contrast, controlled-atmosphere drying produced data points closely aligned with the trend line because the operating conditions remained stable.

As a continuation, Nadhari et al., (2017) reported the drying kinetics of OPT blocks using an isothermal convective dryer. The OPT blocks were prepared with dimensions of 43.91 ± 0.37 mm × 19.73 ± 1.06 mm × 15.82 ± 2.16 mm. The dryer was set to maintain a relative humidity of 80 ± 2% and isothermal temperatures of 25°C, 30°C, 35°C, 40°C, and 45°C. This temperature range was used to investigate the drying behavior of OPT blocks while maintaining their chemical characteristics. The drying kinetics of OPT blocks were analyzed using six models: logarithmic, Wang-Singh, page, midilli-kucuk, Henderson-Pabis, Newton, and the Nadhari model. Each model's appropriateness was tested using SSE, R2, χ^2, and RMSE metrics. The results showed that the Nadhari model best matched the drying behavior of OPT blocks throughout isothermal temperatures. Furthermore, higher isothermal drying temperatures resulted in increased drying rate, porosity, shrinkage, and effective moisture diffusivity (D_{eff}). At 25-40°C, the Deff values for OPT blocks varied between 1.22 × 10-6 and 2.39 × 10-6 m^2/h. The activation energy for drying OPT blocks was found to be 35.56 kJ/mol, which is a rather high figure indicating the OPT blocks' highly bonded moisture content.

1.5. The Pyrolysis of Oil Palm Fronds

Maulina and Silia (2018) evaluated how pyrolysis temperature and residence time affected the properties of bio-oil generated from OPF, such as yield, phenolic content, total acidity, and pH. The pyrolysis temperatures used were 150°C, 200°C, and 250°C, with residence periods of 60, 90, and 120 minutes. The results showed that the maximum bio-oil output, up to 43.47%, was achieved at 150°C with a 2-hour residence duration. The bio-oil with the highest phenolic content was obtained at 250°C and 1 hour, whereas the highest total acidity (11.23%) was observed for bio-oil produced at 150°C for 1 hour. All bio-oil samples obtained under various operating conditions exhibited pH values ranging from 3.1 to 3.8.

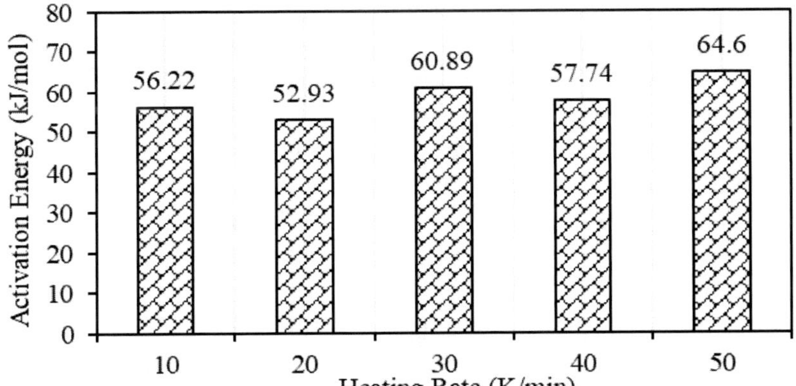

Figure 32. Activation energy of OPF pyrolysis (Soon et al., 2016).

Researchers have also focused on determining the pyrolysis kinetic parameters of OPF. Soon et al., (2016) used thermogravimetric analysis (TGA) to study the pyrolysis kinetics of OPF, specifically examining thermal degradation behavior and kinetic analysis of the energy released during pyrolysis. The pyrolysis experiments were conducted under non-isothermal conditions with heating rates of 10, 20, 30, 40, and 50 K/min over a temperature range of 323–1173 K. Further analysis reveals a trend in the changes of the activation energy (E_a) of OPF pyrolysis ranging from 52.93 to 64.6 kJ/mol, as shown in Figure 32. Thermogravimetric curves indicated a slow mass loss from room temperature to 500 K, corresponding to moisture evaporation. After moisture removal, a significant mass loss occurred in the range of 500–650 K, representing the main devolatilization phase in which volatile matter and carbon were released. Mass loss continued at a slower rate between 650–1050 K and stayed basically steady since 1050–1200 K. At the conclusion of the experiments, the residual solid ranged from 8.0 to 15.9 wt%, depending on the heating rate.

1.6. The Pyrolysis of Oil Palm Trunks

Umar et al., (2020) characterized OPT biomass using CHNSO analysis and X-ray fluorescence (XRF). The results indicated that OPT contains substantial amounts of carbon, hydrogen, and oxygen, with values of 43.35%, 6.26%, and 49.18%, respectively. Nitrogen and sulfur contents were relatively low, at 0.76% and 0.45%, reflecting good fuel quality. XRF analysis revealed CaO and K_2O are the most common oxides in OPT. Although SiO_2 was detected, which can potentially induce agglomeration during thermochemical conversion, its negative effects can be mitigated due to the presence of CaO and Fe_2O_3. The calorific value of OPT was also high, at 17.41 MJ/kg. Overall, OPT exhibits favorable properties for thermochemical conversion, including gasification, combustion, and pyrolysis, due to its adequate C, H, and O content, high calorific value, and low susceptibility to sintering and agglomeration.

Nudri et al., (2020) assessed the suitability of OPT biochar as a solid fuel based on slagging and fouling tendencies, bulk density, proximal and ultimate composition, high calorific value, energy densification ratio, and mass and energy yield. The biochar was created in a top-lit updraft reactor with a maximum temperature of 550°C and a constant airflow rate of 4.63 L/min. The results revealed that the resulting biochar had a bulk density of 87.7 kg/m^3,

4.63% lower than the raw biomass, owing to increased porosity after pyrolysis. The biochar composition resembled lignite, with a 1.87-fold decrease in volatile matter, while fixed carbon and HHV increased to 30.9% and 19.6 MJ/kg, respectively. Nevertheless, the energy yield of the biochar remained low (37.6 ± 1.9%) due to low mass yield. Pyrolysis also produced biochar with a higher ash content than the raw biomass, potentially exacerbating slagging and fouling in furnaces and boilers. These findings highlight the importance of post-processing OPT biochar before co-firing with coal, including demineralization, lowering pyrolysis temperature, and briquetting.

Chantanumat et al., (2023) reported the characteristics of OPT pyrolysis products at 600°C for 1 hour in a fixed-bed reactor. Thermogravimetric study found three major pyrolysis stages: water evaporation (50-100°C), primary devolatilization (200-400°C), and slow carbon decomposition over 400°C, with additional decomposition occurring between 400 and 470°C. The pyrolysis produced two major products: bio-oil and biochar. The bio-oil was primarily composed of aromatic hydrocarbons, including 2,6-dimethoxyphenol, phenol, and n-hexadecanoic acid, and had a calorific value of 27.29 MJ/kg. The produced biochar included 66.30% carbon, 1.56% hydrogen, and 18.22% oxygen, and had a calorific value of 22.64 MJ/kg. Textural investigation showed a specific surface area of 4.43 m^2/g and an average pore diameter of 1.01 Å. This biochar is useful as a fuel, adsorbent, or catalytic support, especially following activation or surface modification.

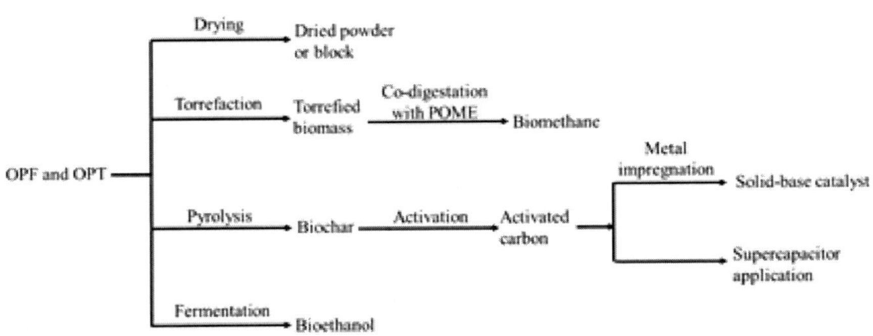

Figure 33. General schematic route for OPF and OPT processing.

Raksodewanto et al., (2024) investigated the utilization of fixed-bed reactors for OPT pyrolysis. Because of the high ash concentration of oil palm biomass, pyrolysis is required to make bio-oil. Bio-oil with a high carbon content and low oxygen concentration is a viable alternative crude oil. OPT was shown to have up to 55.8% carbon. In this investigation, pyrolysis was carried out at a rate of 3°C/min to achieve 600°C, with a reaction time of 1.5 hours. The outer trunk fraction of OPT with a particle size of 40 mesh produced the maximum bio-oil yield (36.8%). Similar patterns were observed when the biomass size was increased to 20 mesh, with the outer trunk fraction yielding the highest bio-oil content (35.7%).

Table 6. Comparison of processing method, product output, and constraints on the oil palm fronds and trunks processing

Processing Method	Product Output	Results	Constraints
Pyrolysis of OPF to obtain activated carbon, surface modification using acetic acid, followed by impregnation using ZrO2	Activated carbon-based ZrO2 catalyst	The catalyst has a surface area of 231.22 m2/g with average pore size of 4.09 nm. Biodiesel yield of 97.36%, with the catalyst reusability about 5 cycles.	Possibility of waste water formation after the chemical-based surface modification
Torrefaction of OPF for co-substrate in POME co-digestation	Torrefied OPF	Torrefied OPF contains hydroxyl, carbonyl, and ether groups. Torrefied OPF can enhance the methane yield up to 11.36 L CH4/g VS.	Biomass torrefaction must begin with a pre-drying stage to reduce the raw material's water content, thereby increasing operational costs
Fermentation	Bioethanol	Low-cost route for energy generation	The quick loss of fermentable carbohydrates during storage
Pyrolysis	Activated carbon for supercapacitor application	The activated carbon has a large surface area (2054.6 m2/g) by using NaOH as an activator. The fabricated supercapacitor has a specific capacitance up to 490.0 F/g, with energy density of 68.1 Wh/kg and power density of 2083.3 W/kg	Possibility of waste water formation after the chemical-based surface modification
Pyrolysis	OPT-derived biochar	The fixed carbon of 30.9% and HHV of 19.6 MJ/kg	Energy yield remained low

Overall, various processing methods can be applied to manage OPF and OPT biomass, both biologically and thermally, as highlighted in Figure 33 and Table 6. Biologycal processes can produce bioethanol as an alternative fuel. However, the problem of rapid loss of fermentable carbohydrates during storage requires serious attention to maintain the stability and quality of bioethanol. Thermal processes such as pyrolysis can produce biochar with a high calorific value, but the possibility of low energy yield remains, as thermal processing can yield a low product mass. Furthermore, functionalization with chemical compounds can improve biochar structure, enabling broader applications. However, managing the resulting waste water remains a classic challenge, especially on an industrial scale. Therefore, in-depth and comprehensive studies are needed to maximize the potential of OPF and OPT, both in terms of product quality and process efficiency.

Conclusion

This chapter examines the potential of oil palm fronds (OPF) and oil palm trunk (OPT) derived from the palm oil plantation sector as biomaterials for a variety of applications. OPF has been utilized for the synthesis of nanocatalysts for biodiesel production, as well as processed through torrefaction, bioethanol production, and supercapacitor development for electrical applications. Both OPF and OPT were explored through a drying approach to produce high-quality biomaterial. Several drying equipment was reviewed, followed by drying kinetic modelling, moisture effective diffusivity coefficient, as well as drying energy activation analysis. Other methods, such as pyrolysis, also demonstrates promising potential for OPF and OPT valorization. Pyrolysis approaches, as a thermochemical conversion, can produce valuable products such as bio-oil or bio-char. This chapter also summarized the kinetic modeling of OPF and OPT pyrolysis using thermogravimetric data. Overall, the valorization of OPF and OPT represents a strategic pathway to support renewable energy production, the development of functional materials, and the enhancement of the value of palm oil industry residues

Chapter 7

The Potential of Oil Palm Leaves

Abstract

Oil palm leaves (OPL) possess significant potential as functional materials in the renewable chemical industry, particularly as adsorbents and catalysts for biodiesel production. As adsorbents, OPL have been successfully modified using both acid and base activators and applied for the removal of pollutants such as tetracycline and the dye Rhodamine B. The developed OPL–based adsorbents have demonstrated substantial adsorption capacities. On the other hand, OPL can also be utilized as catalysts through various processing routes. The calcination–impregnation pathway enables the production of effective silica-based catalysts, whereas hydrothermal carbonization produces hydrochar catalysts with promising catalytic properties. As a result, valorizing OPL not only helps with biomass waste management but also gives a long-term alternative for functional ingredients in biodiesel manufacturing processes.

1. Introduction

The use of oil palm leaves (OPL) has become an important topic in efforts to reduce environmental impact and improve efficiency in the palm oil industry. OPL, a byproduct of oil palm pruning, is commonly regarded as a material of little to no economic use. However, along with the growing awareness of environmental and sustainability issues, research has increasingly focused on developing various technologies to convert OPL into high-value products.

One of the primary applications of OPL is as a feedstock for bioenergy, such as bioethanol and biogas. OPL contains a high concentration of cellulose, which can be converted into bioethanol via fermentation processes. Furthermore, its lignin content makes it a viable raw material for the manufacture of briquettes or other solid fuels, which can be used as renewable energy sources. The use of OPL in energy production helps to reduce carbon emissions and dependence on fossil fuel supplies.

Beyond energy applications, OPL also holds potential for organic fertilizer production. Naturally decomposed OPL is rich in nutrients beneficial to soil, making it suitable as a raw material for nutrient-enriched compost. With proper treatment, compost derived from OPL can improve soil fertility, reduce the reliance on chemical fertilizers, and support sustainable agriculture. In the industrial sector, OPL fibers can be employed as raw materials for composite products such as building materials, particle boards, and other materials. The fibers possess favorable tensile strength, making them suitable for manufacturing lightweight yet durable composites, thereby offering an opportunity to replace more expensive and less eco-friendly synthetic materials.

Furthermore, OPL can be used in paper and package manufacturing. OPL fibers can be processed to provide pulp for recycled paper or environmentally friendly packaging solutions. This not only helps to reduce waste, but it also reduces reliance on wood as the principal raw source in the paper industry. Thus, OPL usage has tremendous potential to promote the ideas of a circular economy, in which waste from one manufacturing process is used as raw material for another, reducing waste generation and increasing sustainability in the palm oil industry as a whole. In accordance with technological advancements, the utilization of OPL can serve as an intelligent solution to mitigate environmental impacts while increasing value-added opportunities within the palm oil sector (Bina Sawit Makmur, 2024).

1.1. Oil Palm Leaves-Based Adsorbents

Researchers have investigated the conversion of OPL waste into a sustainable raw material for activated carbon (AC) production, owing to its low or even negligible cost and abundant precursor content. Among various chemical activation agents, KOH has been shown to produce highly microporous AC while promoting the formation of -OH functional groups on the carbon surface. KOH's potential stems from its ability to form intercalation compounds with carbon. Activation produces K_2O, which is then reduced to metallic K by carbon. This process causes carbon gasification and CO_2 release, which promotes pore formation on the AC surface.

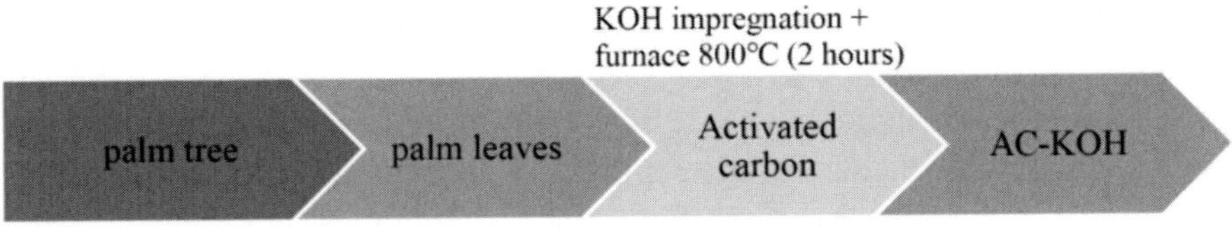

Figure 34. Schematic representation of KOH-activated OPL-based adsorbent production (Islam et al., 2024).

Islam et al., (2024) investigated this approach and reported the development of a KOH-activated carbon-based adsorbent derived from OPL waste. In their study, the adsorbent was prepared from OPL waste through a two-step process. First, OPL was impregnated with 1 M KOH solution and subjected to sonication for 4 hours. The impregnated OPL was then carbonized at 800°C under a nitrogen atmosphere, followed by washing with ethanol and HCl to obtain the desired activated carbon. The overall procedure is illustrated in Figure 34. The resulting activated carbon, referred to as KAC, was comprehensively characterized using several analytical techniques, including XRD, FTIR, SEM, BET, and pH_{pzc} measurements. Subsequently, KAC was evaluated as an adsorbent to remove the emerging contaminant tetracycline (TC) from aqueous media (Islam et al., 2024).

Adsorption studies found that TC removal was significantly dependent on solution pH, with greatest removal occurring between pH 4 and 7. The adsorption kinetics followed a pseudo-second-order model, implying a chemisorption process. The Langmuir isotherm model accurately fitted the adsorption data, with a maximum adsorption capacity of 132.94 mg/g at 25°C. Thermodynamic research revealed that the adsorption process was favorable and endothermic, with a rise in disorder at the solid-liquid interface. FTIR study revealed that TC adsorption on KAC is primarily driven by π-π interactions, hydrogen bonding, electrostatic interactions, and pore filling. Further DFT calculations validated the experimental findings, demonstrating the existence of strong hydrogen bonding interactions between TC and the KAC surface. Recycling studies demonstrated that KAC could be effectively regenerated and reused for TC removal from real wastewater, highlighting its practical application potential. This comprehensive investigation of adsorption mechanisms, combining experimental and computational approaches, provides valuable insights into the interactions between TC and biomass-derived activated carbon. The study underscores the potential of sustainable and environmentally friendly adsorbents for the treatment of water and wastewater contaminated with pharmaceuticals and other emerging pollutants (Islam et al., 2024).

Furthermore, Rahmani et al., (2024) investigated the preparation and use of OPL-based adsorbents for the removal of methylene blue from aqueous solutions. The adsorbent was created using a simple two-step procedure that included carbonization and acid modification of OPL. Figure 35 shows the process of preparing activated carbon. Based on XRD analysis, the adsorbent exhibited an amorphous structure with crystalline and graphitic characteristics. FTIR tests indicated the presence of functional groups including O-H, C-H, C=C, and C-O on the adsorbent surface, indicating its potential for adsorption. The solution pH had a substantial impact on methylene blue removal, with better removal effectiveness seen under alkaline circumstances due to increased electrostatic interaction between the negatively charged adsorbent surface and the cationic form of methylene blue. Increasing the pH from 2 to 12 enhanced the removal efficiency from 28.1% to 85.4%. This is due to the interaction between the adsorbent's surface charge (pHzpc = 7.36) and the dominant cationic form of methylene blue at higher pH values.

Adsorption capacity and removal efficiency decreased with increasing starting dye concentration, owing to the restricted number of active sites on the adsorbent surface at higher concentrations. The removal efficiency increased from 38.5% to 78.3% as the contact time increased from 15 to 180 minutes, with most of the removal occurring within the first 60 minutes. Increasing the adsorbent dose from 25 to 100 mg also improved the removal efficiency from 43.9% to 75.5% but further increases in dose had only a minimal effect. The adsorbent's adsorption capacity decreased from 175.6 mg/g to 32.9 mg/g as the adsorbent dose increased from 25 to 250 mg. However, the adsorption capacity increased from 24.9 mg/g to 96.4 mg/g as the initial dye concentration increased from 25 to 200 mg/L.

The adsorption kinetics followed the pseudo-second-order model, suggesting that chemisorption was the dominant mechanism for methylene blue removal. Furthermore, the adsorption isotherm data were best described by the Freundlich model, indicating heterogeneous adsorption on the adsorbent surface. The results revealed that the highest removal efficiency (99.6%) was achieved at an initial dye concentration of 25 mg/L, an adsorbent dose of 100 mg, and a reaction time of 60 minutes under near-neutral pH conditions. Evaluation of adsorption isotherms and kinetics confirmed that methylene blue adsorption onto the adsorbent was a heterogeneous chemical process following pseudo-second-order kinetics. The potential application of OPL-derived adsorbents proved highly effective in removing dye compounds from industrial wastewater under nearly neutral conditions. The simplicity of the adsorbent preparation process and its high dye removal efficiency make this approach a promising and cost-effective solution to mitigate environmental pollution caused by textile industry effluents (Rahmani et al., 2024).

In liquid-phase adsorption, Khan et al., (2024) synthesized OPL powder as an adsorbent for Rhodamine-B (RhB) from pollutant solutions. The adsorbent was prepared by washing, drying, and grinding OPL into powder with particle sizes below 0.25 mm, as illustrated in Figure 36. Batch adsorption tests were conducted to determine the effects of contact time, adsorbent dosage, initial RhB concentration, temperature, and pH on adsorption capacity. Regeneration studies demonstrated that OPL powder could be recycled through desorption using HCl and reused in subsequent adsorption experiments.

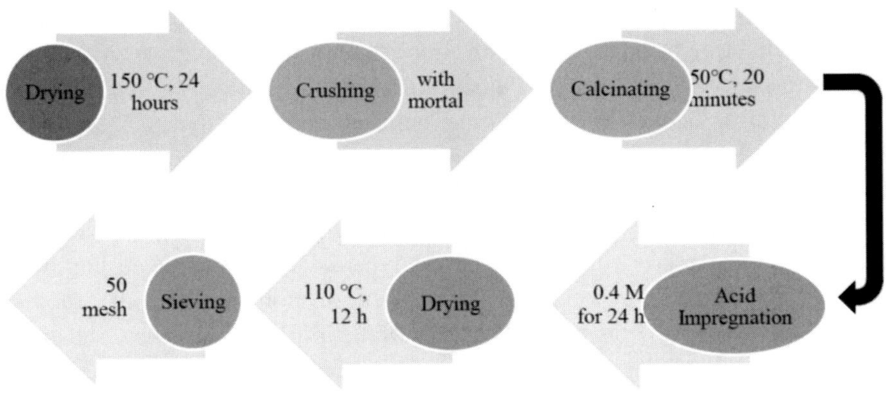

Figure 35. Adsorbent preparation from OPL waste (Rahmani et al., 2024).

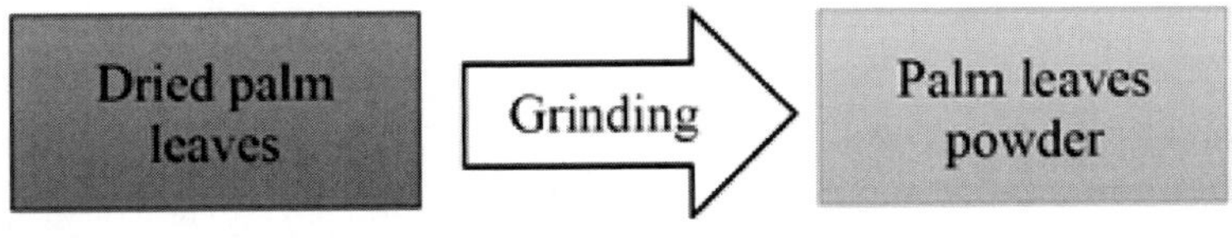

Figure 36. Preparation of OPL powder (Khan et al, 2024).

FTIR analysis revealed the presence of functional groups such as carboxyl, aldehyde, ketone, ester, hydroxyl, and aromatic groups on the surface of palm leaf powder. SEM analysis showed that the surface of the palm leaf powder was rough, which enhances the adsorption of dye molecules. BET analysis indicated a surface area of 3.6 m^2/g for the palm leaf powder, with mesoporous characteristics and an average pore diameter of 14.7 nm. XRD analysis confirmed the existence of cellulose and hemicellulose in the palm leaf powder, and the adsorbent remained stable after adsorbing RhB. The RhB adsorption capacity increased with higher initial dye concentrations and longer contact times but decreased with higher adsorbent doses, temperature, and pH. The reduction in adsorption capacity with increasing temperature suggests that the RhB adsorption process on palm leaf powder is exothermic. The high adsorption capacity at a low pH (pH 3) is due to the protonation of surface functional groups, facilitating the diffusion of monomeric RhB molecules and smaller molecules into the adsorbent pores.

The RhB adsorption on palm leaf powder follows the Langmuir isotherm model, indicating a monolayer adsorption process. The pseudo-second-order kinetic model best describes the adsorption kinetics, indicating that the rate-limiting step is chemical adsorption. Thermodynamic analysis showed that RhB adsorption is an exothermic and non-spontaneous process. The palm leaf powder was regenerated using 0.1 M HCl, and the adsorption efficiency remained above 61% after five adsorption-desorption cycles, demonstrating the potential for adsorbent reuse. Thermodynamic studies revealed that RhB adsorption onto OPL powder was an exothermic, non-spontaneous process. The negative enthalpy change ($\Delta H°$ = -16.95 kJ/mol) and positive Gibbs free energy change ($\Delta G°$) indicate that the adsorption process is exothermic and non-spontaneous.

The adsorption behavior of RhB onto OPL powder highlights its potential as a cost-effective and environmentally friendly adsorbent for the removal of this cationic dye from aqueous solutions. The unique insights from this study demonstrate that the utilization of readily available and sustainable plant-based materials, such as oil palm leaves, can serve as a promising approach for wastewater treatment containing dyes, this contributes to the advancement of effective and environmentally friendly adsorption technologies (Khan et al., 2024).

Table 7. Utilization of OPL as a silica-based catalyst for biodiesel production

Article Title	Method	Main Findings	Catalyst Characteristics
The Potential of Catalyst Based on Palm Leaves in Biodiesel Production as Part of Palm Oil Life Cycle Assessment (Manurung et al., 2024)	Silica was obtained from OPL ash via calcination at 700°C for 3 hours, followed by washing with 1 M HCl solution. Sodium silicate catalysts were synthesized with varying molar ratios of NaOH:SiO2 (1:1, 1.5:1, 2:1, and 2.5:1). The transesterification reaction of RBDPO with methanol was conducted using 2 wt% catalyst at 65°C for 2 hours	The optimum catalyst performance was achieved at a NaOH:SiO2 molar ratio of 1.5:1, yielding 74.485% biodiesel with an ester content of 97.293%. The catalyst showed mesoporous properties, with a surface area of 6.43 m2/g and an average pore size of 5.3 nm. It could be reused up to three times, with the biodiesel production dropping to 62.388%. After catalyst regeneration, the yield further declined to 45.2%	The sodium silicate catalyst derived from OPL ash exhibited the presence of Si–O–Na functional groups as confirmed by FTIR analysis. The catalyst possessed a mesoporous structure with a surface area of 6.43 m2/g and could be regenerated using methanol
Effect of solid base catalyst based silica from palm leaves in biodiesel production (Manurung et al., 2025)	The preparation method was similar, involving calcination of OPL at 750°C for 3 hours, followed by impregnation with NaOH, KOH, and LiOH, and subsequent transesterification CPO	The most effective catalyst loading was 3 wt%, with a reaction period of 60 minutes at a temperature of 60-65°C and an oil-to-methanol molar ratio of 9:1, creating biodiesel with a yield of up to 97.35%.	The mesoporous sodium silicate catalyst exhibited an amorphous structure and maintained catalytic activity up to the third cycle without regeneration
Preparation and characterization of lithium, sodium, and potassium silicate from palm leaf as a potential solid base catalyst in developed biodiesel production (Manurung et al., 2024)	OPL were calcined at 700°C for 3 hours, washed with 1 M HCl, and dried at 105°C for 2 hours. A mixed solution of LiOH, KOH, and NaOH with a weight ratio of 1:2 (w/w) was stirred at 350 rpm and heated at 90°C for 2 hours, followed by drying at 150°C for two hours and calcination at 400°C for an hour.	The biodiesel yield reached 87.50% after only 60 minutes at 60°C, using a catalyst loading of 3% and an oil-to-methanol molar ratio of 1:12	FTIR analysis revealed the presence of Si–OH functional groups in the catalysts (Li2SiO3, Na2SiO3, and K_2SiO_3). The surface areas of Li2SiO3, Na2SiO3, and K_2SiO_3 were 8.250, 8.864, and 8.439 m^2/g, respectively. Among them, Li2SiO3 exhibited the highest basicity, with a value of 2.05.

1.2. The Utilization of Oil Palm Leaves as Green Catalyst

OPL is rich in organic and mineral compounds which, through specific chemical and thermal processes, can be transformed into carbon- and mineral-based heterogeneous catalysts that are effective in accelerating transesterification reactions. The use of OPL as a catalyst not only provides a unique method for recycling solid waste, but also contributes to the development of ecologically favorable, low-cost, and locally accessible catalytic technologies. Consequently, this approach represents an important strategy to establish more efficient, sustainable, and cost-effective biodiesel production processes, supporting national renewable energy agendas while increasing the added value of the palm oil industry. Table 7. summarizes various investigations on silica-based catalysts generated from OPL.

All three studies summarized in Table 7 employed calcination of OPL at approximately 700°C for 3 h as a key step to obtain silica (SiO_2) from the biomass. This was followed by washing with HCl solution to remove contaminants and improve the purity of the silica. In the most recent study, the introduction of KOH and LiOH as alternative bases provided additional complexity and opportunities to explore alkali-metal-based catalysts beyond sodium.

Overall, the synthesis of sodium silicate catalysts from OPL ash has shown to be a viable way for creating biomass-derived heterogeneous catalysts with adequate activity in the transesterification of palm oil into biodiesel. The molar ratio of $NaOH:SiO_2$ at 1.5:1 was identified as the optimum condition. Variations in biodiesel yield and catalytic performance may arise from differences in experimental details such as catalyst loading, washing procedure, and the choice of base (NaOH alone or combined with KOH and LiOH). Mesoporous surface area and the existence of Si-O-Na functional groups were discovered as significant properties that directly correlate with catalytic activity. Although these catalysts can be reused, the gradual loss of activity highlights the importance of developing sustainable regeneration strategies for industrial applications. Further improvements may be achieved through surface modifications or metal doping to enhance catalyst durability.

According to Manurung et al., (2024), the FTIR analysis of palm oil leaf ash and lithium, sodium, and potassium silicate catalysts from combustion showed that in all samples, Si-OH functional groups appeared at wave numbers 3100-3400 cm^{-1}. The hydroxyl group present in these samples corresponds to water vapor bound to the Si element. Furthermore, no peaks were detected in the 2500-3000 cm^{-1} range in any of the samples, indicating the absence of organic compounds due to the silica purification process. Additionally, the Si-OH group in its flexible vibration mode was found in the palm oil leaf ash at 1651 cm^{-1}, while in the catalysts, it was observed in the 1630-1660 cm^{-1} range. The OH group in this region represents a type of hydroxyl group that can vibrate due to thermal treatment on its structure, which is strong and rigid.

XRD analysis revealed that the lithium, sodium, and potassium silica catalysts obtained from calcination produced peaks in the 2θ range of 35°-50°, with the strongest peaks between 37°-39°. This indicates the formation of a crystalline phase due to the calcination process. It can be concluded that the impregnation of LiOH, NaOH, and KOH on palm oil leaf silica ash was successfully carried out.

BET and BJH analyses showed that the catalyst samples derived from palm leaves displayed an adsorption-desorption graph characteristic of type II, which is typical for non-porous or macroporous materials based on IUPAC classification. Morphology of the solid base catalyst after calcination had a structure with a flat surface area and smaller pore volume. This is because metal ions of Li, Na, and K were evenly distributed on the surface of the palm leaf silica ash, reducing diffusion space.

After calcination, the solid base catalyst was analyzed using SEM-EDS to determine the percentage of alkali metal impregnated and distributed on the silica surface. Higher alkali content indicates successful alkali impregnation. SEM observations showed that the catalyst morphology appeared flatter than that of oil palm leaf ash. This is because Li, Na, and K metals have been distributed on the surface of oil palm leaf ash. This finding from the morphological analysis is consistent with the results of the textural analysis using BET-BJH, where the alkali metal-impregnated catalyst had a smaller diffusion space compared to oil palm leaf ash due to the alkali metal impregnation. Overall, the morphological characteristics and elemental composition of the catalyst, along with the identification of functional groups and crystallinity, confirmed the successful process of alkali metal impregnation on the surface of oil palm leaf ash as a matrix. Furthermore, alkalinity analysis results showed that the use of NaOH, LiOH, and KOH alkali solutions is potential to create a base catalyst for biodiesel production due to the reactive and alkaline nature of these solutions. The alkalinity of the catalyst influences the conversion achieved. In this study, sodium and potassium silicates were used as catalysts in the palm oil transesterification process. The reaction conditions for this process were a catalyst weight of 3% of the oil, a methanol-to-oil molar ratio of 12:1, a temperature of 60°C, and a reaction time of 1 hour. The Li_2SiO_3 catalyst resulted in a lower biodiesel yield due to its excessively high alkalinity compared to the other catalysts (Manurung et al., 2024).

Zuhri et al., (2024) also carried out a more detailed study on the synthesis of palm leaves with sodium impregnation. The biodiesel yield increased from 61.514% at a $NaOH:SiO_2$ ratio of 1:1 to 72.468% at a 1.5:1 ratio. However, further increasing the ratio to 2:1 and 2.5:1 led to a decrease in the biodiesel yield, likely due to the overuse of the catalyst, which caused issues with viscosity and saponification. Biodiesel produced at $NaOH:SiO_2$ ratios of 1:1 and 1.5:1 met the Indonesian National Standard (SNI), with a methyl ester content of ≥96.5%. The sodium silicate catalyst could be reused up to three cycles without regeneration, yielding 40.018% biodiesel in the third cycle. The reduction in yield with reuse was likely due to contamination from reaction products and a decrease in the number of basic sites on the catalyst.

In addition to silica-based catalysts, the use of OPL as hydrochar for biodiesel generation has been documented. Aliyu et al., (2024) studied a solid acid catalyst obtained from OPL hydrochar for the effective generation of biodiesel from palm fatty acid distillate (PFAD). The catalyst was prepared via hydrothermal carbonization (HTC) of OPL, which was then activated with phosphoric acid and sulfonated with sulfuric acid. The resulting catalyst, designated PLAHC-S, was characterized using various analytical techniques and applied in PFAD esterification to produce biodiesel. Reaction optimization was carried out using a one-variable-at-a-time (OVAT) approach, and the optimal conditions were found at a catalyst loading of 4 wt%, reaction temperature of 353.15 K, reaction time of 4 h, and methanol-to-PFAD molar ratio of 18:1. Under these conditions, the highest biodiesel yield of 93.56% was obtained. The catalyst was also highly reusable, with biodiesel yields of more than 80% over five successive cycles. The manufactured biodiesel satisfied the physicochemical specifications of ASTM D6751 standards.

The process exhibited pseudo-first-order kinetics, with an activation energy of 23.70 kJ/mol. Thermodynamic characteristics showed that the esterification process was endothermic and not spontaneous. This work presented a novel technique to biodiesel production while also helping to valorize OPL trash. The use of OPL waste as a precursor for catalyst production creates a sustainable avenue for biomass utilization. The hydrothermal carbonization method employed in this study offers an energy-efficient alternative compared to conventional pyrolysis or carbonization techniques. Moreover, the high catalytic activity and reusability of PLAHC-

S, together with its ability to convert low-cost, high–free fatty acid feedstocks such as PFAD into high-quality biodiesel, demonstrate significant potential for reducing production costs. These findings further highlight the substantial potential of OPL waste valorization (Aliyu et al., 2024).

1.3. The Utilization of Oil Palm Leaves in Drug Delivery and Composite Applications

The use of oil palm leaves has also been explored as a material for drug delivery systems, sensors, and composite formation, as investigated by Ajayi et al., (2025). This research focuses on the synthesis and characterization of mesoporous silica nanoparticles (MSNs) utilizing oil palm leaves (OPL) as a silica source. The impact of cetyltrimethylammonium bromide (CTAB) surfactant concentration on the physical properties of MSN was studied using optical microscopy. Silica was extracted from the ash of OPL by initially pretreating the leaves with HCl. This pretreatment process removes contaminants from the OPL ash. Neutralizing the silica solution with H_2SO_4 to near-neutral pH caused precipitation and separation of the silica nanoparticles. The resulting precipitate was rinsed with distilled water to eliminate water-soluble impurities, followed by hydrolysis with NaOH and precipitation using H2SO4. The silica nanoparticles (SNL) were then used to synthesize mesoporous silica nanoparticles (MSN) at various CTAB concentrations (1.0 g, 1.5 g, 2.0 g, and 2.5 g) in a 1-propanol-water solvent mixture. The CTAB concentration was progressively increased to observe its influence on particle morphology, while maintaining a constant ratio of 1-propanol to water (1:1) and temperature. The solubility of CTAB in the 1-propanol-water mixture resulted in ion and micelle formation, which interacted with the silica nanoparticles, weakening the inter-molecular 3D network, allowing CTAB to penetrate the particle interior. The removal of the template from the mesostructure was achieved through calcination at 550°C, preventing the potential collapse or damage of particles that could occur at temperatures above 600°C.

FTIR and EDX analyses confirmed the successful synthesis of MSN. Optical microscopy showed that the concentration of CTAB had a significant effect on the size, morphology, shape, and surface area of MSN. At a CTAB concentration of 1.5 g, MSN (MSL-15) exhibited large pores, with an average surface area ranging from 50-400 $\mu m^2/g$ and particle size between 50-100 µm. This process demonstrates that the CTAB concentration, the length of the alcohol chain, and its interaction with silica nanoparticles control the final characteristics of MSN. The mechanism reveals that during the synthesis process, CTAB orients toward the amorphous silica areas where the 1-propanol-water mixture had been absorbed, while some CTAB adsorbs onto the surface of the particles. At higher concentrations, some counterions of CTAB aggregate at the solution-air and 1-propanol-water interfaces. As the reaction progresses, CTAB rearranges, and the particles that have adsorbed the 1-propanol-water mixture oscillate at the 1-propanol-water interface, with 1-propanol acting as a co-surfactant. During calcination, the template structure folds into different shapes, and the combustion of the absorbed template creates pore sizes depending on the initial agglomeration and condensation orientations. The final shape of the product is influenced by the structural arrangement, burning rate, type of organic gases, and the degree of particle collapse after the template is removed.

The reaction for the formation of silica nanoparticles can be seen in the following reactions, which involve two reaction processes. The first reaction illustrates how a base interacts with silica to form a soluble silicate compound. Then, in the second reaction, sodium metasilicate (Na_2SiO_3) obtained from the first reaction is reacted with hydrochloric acid (HCl). This process results in the formation of silica (SiO_2) in its solid form or as silica nanoparticles, along with sodium chloride (NaCl) that dissolves in water and water (H_2O).

$$2NaOH(aq) + SiO_2 \longrightarrow Na_2SiO_3(aq) + H_2O(l)$$
$$Na_2SiO_3(aq) + 2HCl(aq) \longrightarrow SiO_2(s) + 2NaCl(aq) + H_2O(l)$$

The key factors influencing the physical properties of the mesoporous silica nanoparticles (MSNs) synthesized in this study are: (1) CTAB Concentration; the concentration of cetyltrimethylammonium bromide (CTAB) significantly affects the size, morphology, shape, and surface area of the MSN. At a 10x magnification, mesoporous silica prepared in a 1-propanol-water medium exhibited large pores at 1.5 g of CTAB. (2) Solvent System; the 1-propanol-water (1:1) solvent system was used for MSN synthesis. The solubility of CTAB in the co-solvent mixture of 1-propanol and water results in the formation of ions and micelles, which interact with the silica nanoparticles and weaken the inter-molecular 3D network structure, thus enabling CTAB to penetrate the interior of the particles. This solvent system plays a critical role in determining the physical characteristics of the MSN. (3) Reaction Mechanism; the proposed reaction mechanism explains how the CTAB concentration and the 1-propanol-water solvent system affect the formation and characteristics of MSN. The authors describe the reorientation of CTAB, the aggregation of CTAB counterions, and the role of 1-propanol as a co-surfactant, all of which influence the final physical properties of the MSN (Ajayi et al., 2025).

This research offers valuable insights into the synthesis of mesoporous silica nanoparticles from a renewable and abundant natural resource, oil palm leaves. The ability to control the physical properties of MSN by adjusting CTAB concentration demonstrates the potential of this material for various applications, including drug delivery, sensors, and composite formation.

Conclusion

This research highlights the considerable potential of oil palm leaves (OPL) as a sustainable and multifunctional material in the renewable chemical industry, particularly in adsorption, catalysis, and drug delivery. As an adsorbent, OPL-based materials, modified with both acid and base activators, have shown impressive capabilities in removing pollutants, such as tetracycline and Rhodamine B (RhB), from aqueous solutions. The adsorption performance of these materials was significantly influenced by factors like solution pH, contact time, and adsorbent dosage, with high reusability observed even after multiple adsorption-desorption cycles, indicating their practical applicability for environmental remediation.

In the field of catalysis, OPL has proven to be a valuable raw material for creating silica-based catalysts. These catalysts, produced through calcination and impregnation processes, displayed effective catalytic activity in biodiesel production. Notably, sodium silicate catalysts derived from OPL ash provided high biodiesel yields, with the most efficient performance achieved at a $NaOH:SiO_2$ molar ratio of 1.5:1. These catalysts could be reused up to three cycles, although their catalytic activity declined over time, emphasizing the need for sustainable regeneration methods to maintain efficiency. Further research on surface modifications or metal doping could potentially enhance the catalysts' durability and performance.

The synthesis of mesoporous silica nanoparticles (MSNs) from OPL opens up additional possibilities beyond adsorption and catalysis. By controlling the concentration of cetyltrimethylammonium bromide (CTAB) and using a 1-propanol-water solvent system, OPL-derived MSNs exhibited adjustable properties such as pore size, morphology, and surface area, making them suitable for diverse applications in drug delivery, sensors, and composite materials. This ability to tailor the material's properties offers an environmentally friendly and cost-effective alternative to synthetic materials in various industries.

Overall, the valorization of OPL not only provides a solution for sustainable waste management by converting agricultural byproducts into valuable functional materials, but it also contributes to the development of green technologies for biodiesel production, environmental pollution control, and biomedical applications. The results of this study underscore the significant role OPL can play in fostering a circular economy, reducing dependence on fossil fuels, and supporting the sustainable development goals of the palm oil industry. Future research may focus on improving catalyst reusability, enhancing the properties of MSNs for more efficient drug delivery, and exploring new applications in other sectors.

Conclusion

The palm oil industry generates a diverse range of biomass, both from plantations and processing mills. The use of solid palm oil leftovers provides substantial opportunity to assist sustainable industrial growth using a circular economy strategy. Several fractions of palm biomass - including empty fruit bunches (EFB), mesocarp fibers, shells, fronds, trunks, and leaves - have been investigated for value-added products. For example, EFB and mesocarp fibers can be fermented to produce bioethanol, a renewable energy source that could replace fossil fuels and lower the palm oil industry's carbon footprint.

In addition, palm biomass holds great potential as a functional material. Shells, fronds, trunks, and leaves can be transformed into adsorbents for pollutant removal from wastewater, and then into catalysts for esterification and transesterification reactions in biodiesel generation. Various processing routes, including calcination, impregnation, and hydrothermal carbonization, enable the formation of silica-based catalysts and hydrochar with high activity. Pyrolysis of palm biomass also yields bio-oil, syngas, and char with broad energy and material applications, thereby opening pathways for product diversification from biomass residues.

The utilization of palm biomass emphasizes not only technical efficiency but also substantial economic impact. By converting waste into value-added goods, the palm oil industry may cut waste management expenses while also generating new revenue streams. In Indonesia, the world's largest palm oil producer, such methods have the potential to promote the circular economy, drive green technical innovation, and improve the sustainability of palm oil supply chains.

Moreover, biomass utilization contributes to reducing solid waste traditionally disposed of or burned, thus lowering air pollution and greenhouse gas emissions while improving resource-use efficiency. By transforming residues into inputs for other products, the palm oil industry can effectively implement circular economy principles, extend the material life cycle, and create new income streams that reinforce sustainability. Furthermore, renewable energy products derived from biomass, such as bioethanol and bio-oil, help reduce dependence on fossil fuels, while the development of catalysts and adsorbents from biomass promotes green technological innovation across industrial sectors. Biomass valorization can also enhance local economic value, create employment opportunities, and strengthen sustainable development in social, economic, and environmental dimensions. Overall, incorporating palm biomass usage provides a practical step toward a more environmentally friendly, resource-efficient, and sustainable palm oil sector.

References

Abdullah, S. S. S. Syed Abdullah, Bahrin, E. K., Shirai, Y., and Hassan M. A. (2021). Influence of Storage Conditions on Oil Palm Frond Juice as a Renewable Feedstock for Bioethanol Production. *Biomass and Bioenergy*, 150 (April). doi:10.1016/j. biombioe.2021.106101.

Adlie, T. A., Rizal, S., Ali, N., Huzni, S., and Thalib, S. (2019). Characterization of Polymeric Foam Composite Reinforced with Empty Fruit Bunch. *International Conference on Science and Innovated Engineering*. 536. Page 1-9.

Ahmadzadeh, A., and Zakaria, S. (2009). Study on the Water Absorption and Hardness of Phenolated Oil Palm Empty Fruit Bunches (PEFB)-base Biocomposite. *Iranian Polymer Journal,* 18 (3): 247-255.

Ajayi, S. M., Olusanya, S. O., Abimbade, S. F., Olumayede, E. G., Faboya, O. L., Malomo, D., and Akintayo, C. O. (2025). Synthesis of mesoporous silica nanoparticles from oil palm leaves: Mechanism and characterization using optical microscope. *Next Research,* Volume 2, Issue 3, 100522, ISSN 3050-4759, https://doi.org/10.1016/j. nexres.2025.100522.

Akia, M., Yazdani, F., Motaee, E., Han, D., and Arandiyan, H. (2014). A review on conversion of biomass to biofuel by nanocatalysts. *Biofuel research journal,* 1(1), page 16-25.

Aliyu, M., Moser, B. R., Alharthi, F. A., and Rashid, U. (2024). Efficient Production of Biodiesel from Palm Fatty Acid Distillate Using a Novel Hydrochar-Based Solid Acid Catalyst Derived from Palm Leaf Waste. *Process Safety and Environmental Protection,* 187 (May): 1126–39. https://doi.org/10.1016/j.psep.2024.05.040.

Aprianti, N., Kismanto, A., Supriatna, N. K., Yarsono, S., Nainggolan, L. M. T., Purawiardi, R. I., Fariza, O., Ermada, F. J., Zuldian, P., Raksodewanto, A. A., & Alamsyah, R. (2023). Prospect and challenges of producing carbon black from oil palm biomass: A review. *Bioresource Technology Reports,* 23(July). https://doi.org/ 10.1016/j.biteb.2023.101587.

Atminingsih, A., Junaidi, J., Wahyuni, M., and Febrianto, E. B. (2020). Pemanfaatan Tandan Kosong Kelapa Sawit Sebagai Media Tanam Pembibitan Batang Bawah Karet dengan Metode Root Trainer. *Jurnal Agro Industri Perkebunan*, 7(2) page 99-108.

Ausavasukhi, A., and Sooknoi, T. (2025). Hydrothermal Sulfonation of Palm Kernel Shells to Produce a Carbon-Based Solid Acid Catalyst for the Glycerol Etherification. *Diamond and Related Materials,* 154 (March). https://doi.org/10.1016/j.diamond. 2025.112217.

Ayu, R.S., Khalina, A., Harmaen, A. S., Zaman, K., Isma, T., Liu, Q., Ilyas, R. A., and Lee, C. H. (2020). Characterization Study Of Empty Fruit Bunch (EFB) Fibers Reinforcement in Poly(Butylene) Succinate (PBS)/Starch/Glycerol Composite Sheet. *Polymers.* 12(7): 1-13. https://doi.org/10.3390/polym12071571.

Babatunde, E.O., Enomah, S., Akwenuke, O.M., Ibeh, M.A., Okwelum, C.O., Mundu, M.M., Adepoju, P.O., Aki, A.O., Oghenejabor, O.D., Adepoju, T.F., Ifedora, C.O., and Mabel, K. (2025). Preparation and Characterization of Palm Kernel Shell (PKS) Based Biocatalyst for the Transformation of Kernel Oil to Biodiesel. South African *Journal of Chemical Engineering,* 52 (February): 200–210. https://doi.org/10.1016/j.sajce.2025.02.008.

Barbosa, S. & Castillo, L. (2020). Biodegradable Pots for Seedlings. Advanced Applications of Biodegradable Green Composites. *Materials Research Foundations,* Vol. 68, pp 104 137. doi: http://dx.doi.org/10.21741/9781644900659-5.

Basu, P. (2018). *Biomass Gasification, Pyrolysis and Torrefaction: Practical Design and Theory.* Academic Press.

Bicalho, T., Bessou, C., and Pacca, S. A. (2016). Land use change within EU sustainability criteria for biofuels: the case of oil palm expansion in the Brazilian Amazon. *Renew Energy*, 89:588–97.

Bina Sawit Makmur. (2024). *Pemanfaatan Limbah Kelapa Sawit.* www.binasawitmakmur.com. https://binasawitmakmur.com/berita/pemanfaatan-limbah-kelapa-sawit.

Bonechi, C., Consumi, M., Donati, A., Leone, G., Magnani, A., Tamasi, G., & Rossi, C. (2017). Biomass: An overview. *Bioenergy Systems for the Future* (pp. 3–42). https://doi.org/10.1016/B978-0-08-101031-0.00001-6.

Burhani, D., Septevani, A. A., Setiawan, R., Djannah, L. M., and Pratama, M. A. P. (2022). The Effect of Drying Process of Cellulose Nanofiber from Oil Palm Empty Fruit Bunches on Morphology. In *IOP Conference Series: Earth and Environmental Science*, page 1–8.

Callister, W. D. Jr. (2018). *Material Science and Engineering an Introduction.* Departement of Chemical and Biochemical Engineering. Wiley.

Chang, J. S. L., Law, M. C., Chan, Y. S., and Leo, C. P. (2015). Effects of Microwave Heating on Oil Palm Mesocarp. *Chemical Engineering Transactions*, 45: 1633–38. https://doi.org/10.3303/CET1545273.

Chang, J. S.L., Chan, Y.S., Law, M.C., and Leo, C.P. (2017). Comparative Microstructure Study of Oil Palm Fruit Bunch Fibre, Mesocarp and Kernels After Microwave Pre-Treatment. *IOP Conference Series: Materials Science and Engineering,* 217: 1–10. doi:10.1088/1757-899X/217/1/012026.

Chantanumat, Y., Phetwarotai, W., Sangthong, S., Palamanit, A., Abu-Bakar, M. S., Cheirsilp, B., and Phusunti, N. (2023). Characterization of Bio-Oil and Biochar from Slow Pyrolysis of Oil Palm Plantation and Palm Oil Mill Wastes. *Biomass Conversion and Biorefinery,* 13: 13813–25. doi:10.1007/s13399-021-02291-2.

Di Blasi, C. (2008). Modeling chemical and physical processes of wood and biomass pyrolysis. *Progress in energy and combustion science*, 34(1), page 47-90.

Fudholi, A., Sopian, K., Alghoul, M.A., Ruslan, M. H., and Othman, M. Y. (2015). Performances and Improvement Potential of Solar Drying System for Palm Oil Fronds. *Renewable Energy*, 78: 561–65. doi:10.1016/j.renene.2015.01.050.

Gaol, M. R. L. L., Manurung, R., S., Yanthi., Surya, I. and Sitorus, R. (2013). Pembuatan Selulosa Asetat Dari A -Selulosa Tandan Kosong Kelapa Sawit. *Jurnal Teknik Kimia* USU, Vol. 2, No. 3.

Ghazali, N. F., Pahlawi, Q. A., Hanim, M., and Makhtar, N. A. (2017). Enzymatic Hydrolysis of Oil Palm Empty Fruit Bunch using Membrane Reactor. *Chemical Engineering Transaction.* 56: 1543-1548. doi: 10.3303/CET1756258.

Gollakota, A. R. K., Kishore, N., & Gu, S. (2018). A review on hydrothermal liquefaction of biomass. *Renewable and Sustainable Energy Reviews*, 81, 1378-1392.

Greenough, S., Dumont, M. J., and Prasher, S. (2021). The physicochemical properties of biochar and its applicability as a filler in rubber composites: a review. Mater. *Today Commun.*, 29, 102912. https://doi.org/10.1016/j.mtcomm.2021.102912.

Ha, E. Y. W., and Landi, C. D. (1997). Method for Producing Microcrystalline Cellulose. *Patent* US5769934A.

Halim, M., Wahyudi, E., & Putra, I. A. (2019). Pemberian Pupuk NPK dan Kompos Tandan Kosong Kelapa Sawit pada Pertumbuhan Bibit Kelapa Sawit (Elaeis guineensis Jacq) di Pembibitan Awal. Agrinula: *Jurnal Agroteknologi Dan Perkebunan*, 2(1), 9–12.

Hall, M., Bansal, P., Lee, J. H., and Bommarius, J. R. A. S. (2010). Cellulose Crystallinity-a Key Predictor Of The Enzymatic Hydrolylis Rate. *The FEBS Journal*. 277(6) : 1571-1582. https://doi.org/10.1111/j.1742-4658.2010.07585.x.

Hanan, F., Jawaid, M., Paridah, M. T., and Naveen, J. (2020). Characterization of Hybrid Oil Palm Empty Fruit Bunch/Woven Kenaf Fabric-Reinforced Epoxy Composites. *Polymers.* 12(9): 1-13. https://doi.org/10.3390/polym12092052.

Handoko, S., Nurhadi, N., Mujiati, S., and Fitriani, R. (2021). Characterization of Pyrolysis Products of Oil Palm Empty Fruit Bunch. In *IOP Conference Series: Earth and Environmental Science,* 1–6. doi:10.1088/1755-1315/749/1/012041.

Harahap, H., Nasution, H., Manurung, R., Yustira, A., and Rashid, A. A. (2025). Physical Characteristics of Biodegradable Pots from Empty Fruit Bunches (EFB) and Sawdust Waste as Planting Media. *Case Studies in Chemical and Environmental Engineering*, 11 (January). doi:10.1016/j.cscee.2025.101193.

Hasibuan, R., and Daud, W.R.W. (2004). Through Drying of Oil Palm Empty Fruit Bunches (EFB) Fiber Using Superheated Steam. In *Proceedings of the 14th International Drying Symposium,* (IDS 2004), 2027–34.

Hasibuan, R., and Daud, W.R.W. (2009). Quality Changes of Superheated Steam-Dried Fibers from Oil Palm Empty Fruit Bunches. *Drying Technology,* 27 (2): 194–200. doi:10.1080/07373930802603383.

Henriksson, G., Germgard, U., and Lindsrom, M. E. (2024). A review on Chemical Mechanisms of Kraft Pulping. *Nordic Pulp and Paper Research Journal*. 39(3): 1-15.DOI 10.1515/npprj-2023-0015.

Hon, D. N. S, and Shiraishi, N. (2001). *Wood and Cellulosic Chemistry*. Marcel Dekker Inc.

Hossain, M. A., Jewaratnam, J., and Ganesan, P. (2016). Prospect of hydrogen production from oil palm biomass by thermochemical process – A review. *International Journal of Hydrogen Energy.* 41(38): 16637-16655. https://doi.org/10.1016/j.ijhydene.2016.07.104.

Hudayarizka, R., Yorika, R., and Ariani, I. K. (2024). *Limbah Sawit Sebagai Sumber Energi: Inovasi Untuk Masa Depan Berkelanjutan.*

Ibrahim, M. H. W., Mangi, S. A., Zuki, S. S. M., Jaya, R. P., and Supriyatno, D. (2020). Infulence of Palm Oil Biomass Clinker and Empty Fruit Bunch Fibers on Concrete Properties. *IIUM Engineering Journal.* 21(2). https://doi.org/10.31436/iiumej.v21i2.1285.

Ibrahim, N. A., Rashid, U., Hazmi, B., Moser, B. R., Alharthi, F. A., Rokhum, S. L., and Ngamcharussrivichai, C. (2022). Biodiesel Production from Waste Cooking Oil Using Magnetic Bifunctional Calcium and Iron Oxide Nanocatalysts Derived from Empty Fruit Bunch. *Fuel,* 317 (July 2021). doi:10.1016/j.fuel.2022.123525.

Ibrahim, N. A., Rashid, U., Hin, Y., Choong, T., and Yaw, S. (2019). Synthesis of Carbonaceous Solid Acid Magnetic Catalyst from Empty Fruit Bunch for Esteri Fi Cation of Palm Fatty Acid Distillate (PFAD). *Energy Conversion and Management, Volume* 195, 1 September 2019, Pages 480-491.

Indexmundi, (2011). *Palm Oil Production by Country in 1000 MT.* http: www.indexmundi.com. Diakses 18 Juni 2012 dari Indexmundi.

Iskandar, W. M. E., Ong, R. H., Khan, M. M. R., and Toma. (2024). Effect of bio-based nanosilica on the mechanical properties of single empty fruit bunch fibre. *Aip Conf. Proc.* 3014. https://doi.org/10.1063/5.0192522.

Islam, M. A., Nazal, M. K., Akinpelu, A. A., Sajid, M., Alhussain, N. A., Billah, R. E-K., and Bahsis, L. (2024). Novel Activated Carbon Derived from a Sustainable and Low-Cost Palm Leaves Biomass Waste for Tetracycline Removal: Adsorbent Preparation, Adsorption Mechanisms and Real Application. *Diamond and Related Materials*, 147 (April). doi:10.1016/j.diamond.2024.111375.

Jacquet, N., Manjet, G., Vanderghem, C., Delvigne, F., and Richel. (2015). Application of Steam Explosion As Pretreatment on Lignocellulosic Material: A review. *Ind. Eng Chem. Res.* 54(10): 2593-2598. https://doi.org/10.1021/ie503151g.

Janaun, J., Ellis N. (2010). Perspectives on biodiesel as a sustainable fuel. *Renew Sustain Energy* Rev, 14:1312–20.

Jaya, J. D., Elma, M., Sunardi., Nugroho, A. (2022). Physical and Mechanical Properties of Biodegradable Pot Derived from Oil Palm Empty Fruit Bunch and Sodium Alginate. *Brazilian Archives of Biology and Technology.* Vol.65: e22210789.

Kamide, K. (2005). Cellulose and Cellulose Derivatives Molecular Characterization and its Applications. *Elsevier*. doi:10.1016/B978-0-444-82254-3.X5000-0.

Karppil, A., Vuorenpalo, V. M., Robertsen, L., Parviainen, K., Dahl, O., and Vanhatalo. (2010). Process For Producing Microcellulose. *Patent* US10358504B2.

Kaushika, N. D., Reddy, K. S., & Kaushik, K. (2016). Biomass energy and power systems. In *Sustainable energy and the environment: a clean technology approach* (pp. 121-137). Cham: Springer International Publishing.

Khalil, H.P.S., Bhat, I.B., and Sartika, M.Y. (2010). Degradation, Mechano-Physical, And Morphological Properties of Empty Fruit Bunch Reinforced Polyester Composites. *BioResources*, 5(4) page 2278-2296.

Khan, M. I., Almanassra, I. W., Shanableh, A., Atieh, M. A., Manzoor, S., Hayat, M., Besbes, M., Elgharbi, S., Alimi, F., and Jemmali, M. (2024). Utilization of palm leaves powder for removal of Rhodamine-B from an aqueous solution. *Desalination and Water Treatment,* Volume 317, January 2024, 100159.

Lestari, N., Rahmah, N., Novitasari, E and Samsuar. (2022). Performa biopot berbahan dasar limbah organik sebagai wadah pembibitan tanaman pengganti plastik polybag. Agrika: *Jurnal Ilmu-ilmu Pertanian,* 16 (2): 90-105.

Lim, S., Yap, C. Y., Pang, Y. L., and Wong, K. H. (2020). Biodiesel Synthesis from Oil Palm Empty Fruit Bunch Biochar Derived Heterogeneous Solid Catalyst Using 4-Benzenediazonium Sulfonate. *Journal of Hazardous Materials*, 390 (May 2019): 121532. doi:10.1016/j.jhazmat.2019.121532.

Lukova, P., Katsarov, P., and Pilicheva, B. (2023). Polymers, Application of Strach, Cellulose, and Their Derrivatives in The Development of Microparticle Drug-Delivery Systems. *Polymers.* 15(17) : 1-30. https://doi.org/10.3390/polym15173615.

Manurung, R., Hasibuan, R., and Siregar, A.G.A. (2024). Preparation and Characterization of Lithium, Sodium, and Potassium Silicate from Palm Leaf as a Potential Solid Base Catalyst in Developed Biodiesel Production. *Case Studies in Chemical and Environmental Engineering,* 9 (November 2023). doi:10.1016/j.cscee.2023.100543.

Manurung, R., Hasibuan, R., Siregar, A.G.A., Siregar, F.K., and Siregar, N.K. (2025). Effect of Solid Base Catalyst Based Silica from Palm Leaves in Biodiesel Production. *AIP Conference Proceedings*, 3248(1). doi:10.1063/5.0236786.

Manurung, R., Parinduri, S. Z.D.M., Hasibuan, R., Tarigan, B. H., and Siregar, A. G.A. (2023). Synthesis of Nano-CaO Catalyst with SiO2 Matrix Based on Palm Shell Ash as Catalyst Support for One Cycle Developed in the Palm Biodiesel Process. *Case Studies in Chemical and Environmental Engineering*, 7 (March): 100345. doi:10. 1016/j.cscee.2023.100345.

Manurung, R., Tambun, R., Zuhri, R.R.S., Harahap, H., and Siregar, A.G.A. (2024). The Potential of Catalyst Based on Palm Leaves in Biodiesel Production As Part of Palm Oil Life Cycle Assesment. *EUREKA, Physics and Engineering*, (4): 40–51. doi:10.21303/2461-4262.2024.003433.

Mathew, A. P., Oksman. K., and Sain. M. (2005). Mechanical properties of biodegradable composites from poly lactic acid (PLA) and microcrystalline cellulose (MCC). *Journal of Applied Polymer Science.* 97(5): 2014-2025. https://doi.org/10.1002/app. 21779.

Maulina, S., and Silia, F. (2018). Liquid Smoke Characteristics from the Pyrolysis of Oil Palm Fronds. In *IOP Conference Series: Materials Science and Engineering*, 1–5. doi: 10.1088/1757-899X/309/1/012073.

Milicevic, D. P. M., Chan, Y. J., Shi, S., and Gopakumar, S. T. (2025). Effect of Torrefaction of Palm Frond on Methane Yield through Anaerobic Co-Digestion with Palm Oil Mill Effluent. *Journal of Analytical and Applied Pyrolysis*, 192 (August). doi:10.1016/j.jaap.2025.107316.

Mohideen, M. F., Faiz, M., Salleh, H., Zakaria, H., and Raghavan, V. R. (2011). Drying of Oil Palm Frond via Swirling Fluidization Technique. In *Proceedings of the World Congress on Engineering,* page 2375–80.

Muhammad, M., Munawara, E., and Zaki, M. (2022). Experimental Study of CO2 Utilization as a Density Modification Agent for Maximizing Palm Shells and Kernels Separation Efficiency. *South African Journal of Chemical Engineering,* 42 (September): 283–89. doi:10.1016/j.sajce.2022.09.006.

Nadhari, W. N. A. W., Hashim, R., Danish, O., Sualaiman, M., Karim, N. A., Yahaya, A. N. A., and Saari, N. (2017). Isothermal Drying Kinetics of Oil Palm Trunk: Energy and Shrinkage Evaluation. *Environmental Progress & Sustainable Energy*, page 1–9.

Nadhari, W. N. A. W., Hashim, R., Sulaiman, O., and Jumhuri, N. (2014). Drying Kinetics of Oil Palm Trunk Waste in Control Atmosphere and Open Air Convection Drying. *International Journal of Heat and Mass Transfer*, 68: page 14–20. doi:10.1016/j.ijheatmasstransfer.2013.09.009.

Narasimhan, M., Chandrasekaran, M., Govindasamy, S., and Aravamudhan, A. (2021). Heterogeneous Nanocatalysts for Sustainable Biodiesel Production: A Review. *Journal of Environmental Chemical Engineering,* 9(1): 104876. doi:10.1016/j.jece.2020.104876.

Nasution, H., Suherman, P., Kelvin., and Winny. (2020). Mechanical Properties of Microcrystalline Cellulose from Coconut Fiber Reinforced Waste Styrofoam Composite: *The Effect of Compression Molding Temperature*. 1003: 1-6. doi:10.1088/1757-899X/1003/1/012125.

Nasution, H., Yurnaliza., Veronicha., Irmadani. (2017). Preparation and Characterization of Cellulose Microcrystalline (MCC) from Fiber of Empty Fruit Bunch Palm Oil. *IOP Science Material Science and Engineering*. 180: 1-9. doi:10.1088/1757-899X/180/1/012007.

Nguyen, X. (2002). Process for Preparing Microcrystalline Cellulose. *Patent* US20040074615A1.

Nirmala, W., Purwaningrum, P., & Indrawati, D. (2020). Pengaruh Komposisi Sampah Pasar Terhadap Kualitas Kompos Organik Dengan Metode Larva Black Soldier Fly (BSF). In *Prosiding Seminar Nasional Pakar*, page 1-29.

Noorshashillawati, A.B.M, (2007). Synthesis, Characterization And Properties of The New Unsaturated Polyester Resins For Composite Applications. *Master Thesis*, Universiti Sains Malaysia.

Nordin, N. I. A. A., Ariffin, Hidayah., Andou, Y., Hassan, M. A., Shirai, Y., Nishida, H., and Yunus, W. M. Z. W., Karuppuchamy, S., and Ibrahim, N. A. (2013). Modification of Oil Palm Mesocarp Fiber Characteristics Using Superheated Steam Treatment. *Molecules,* 18: 9132–46. doi:10.3390/molecules18089132.

Novita, Lidya, Safni, Emriadi, Freitas, F. A. D., Fauzia, S., and Zein, R. (2024). Enhanced Conversion of Used Palm Cooking Oil to Biodiesel by a Green and Recyclable Palm Kernel Shell Ash-Derived Catalyst: Process Optimization by Response Surface Methodology. *Case Studies in Chemical and Environmental Engineering,* Volume 9, June 2024, 100678.

Novo, L. P., Bras. J., Garcia, A., and Curvelo, A. A. D. S. (2016). A study of the production of cellulose nanocrystals through subcritical water hydrolysis. *Industrial Crops and Products*. 93: 1-8. https://doi.org/10.1016/j.indcrop.2016.01.012.

Nudri, N. A., Bachmann, R. T., Ghani, W. A. W. A. K., Sum, D. N. K., and Azni, A. A. (2020). Characterization of Oil Palm Trunk Biocoal and Its Suitability for Solid Fuel

Applications. *Biomass Conversion and Biorefinery,* 10: 45–55. doi:10.1007/s13399-019-00419-z.

Nussbaumer, T. (2003). Combustion and co-combustion of biomass: fundamentals, technologies, and primary measures for emission reduction. *Energy & fuels,* 17(6), 1510-1521.

Obada, D. O., Kekung, M. O., Levonyan, T., & Norval, G. W. (2023). Palm oil mill derived empty palm fruit bunches as a feed stock for renewable energy applications in Nigeria: A review. *Bioresource Technology Reports,* 24(April), 101666.

Paramudita, I., Aziz, F., Yuliantini, L., Septina, W., & Nursam, N. M. (2026). Valorization of oil palm frond into activated carbon electrodes with enhanced electrochemical performance for supercapacitor. *Journal of Physics and Chemistry of Solids*, 208(July 2025). https://doi.org/10.1016/j.jpcs.2025.113053.

Pennells, J., Godwin, I. D., Amiralian, N., and Martin, D. J. M. (2020). Trends in the Production of Cellulose Nanofibers from Non-Wood Sources. *Cellulose*, 27: 575–93. doi:10.1007/s10570-019-02828-9.

Purba, I. T., Sani, K. Q., Sayekti, N., Ramadhani, S. S., Waluyo, J., Pranolo, S. H. and Kaavessina, M. (2023). Biofilm Fabrication from Cellulose Acetate of Oil Palm Empty Fruit Bunch and Corn Starch as Bio-polybag Material for Eco-friendly Plantation. *IOP Conf. Series: Earth and Environmental Science* 1217 012037.

Purboputro, P. I. (2017). Pengaruh Panjang Serat terhadap Kekuatan Impak Komposit Enceng Gondok dengan Matriks Poliester. Media Mesin: *Majalah Teknik Mesin,* 7(2).

Puspasari, I., Talib, M. Z. M., Daud, W. R. W., and Tasirin, S. M. (2012). Drying Kinetics of Oil Palm Frond Particles in an Agitated Fluidized Bed Dryer. *Drying Technology,* 30: 619–30. doi:10.1080/07373937.2012.654873.

Puspasari, I., Talib, M.Z.M., Daud, W.R.W., and Tasirin, S.M. (2014). Characteristic Drying Curve of Oil Palm Fibers. International Journal on Advanced Science, *Engineering and Information Technology*, 4(1): 20–24. doi:10.18517/ijaseit.4.1.361.

Quah, R. V., Tan, Y. H., Mubarak, N. M, Kansedo, J., Khalid, M., Abdullah, E. C., and Abdullah, M. O. (2020). Magnetic Biochar Derived from Waste Palm Kernel Shell for Biodiesel Production via Sulfonation. *Waste Management*, 118: 626–36. doi:10. 1016/j.wasman.2020.09.016.

Rahayu, D.E., Karnaningroem, N., Altway, A., and Slamet, A. (2021). Utilization of Oil Palm Empty Fruit Bunches Biomass Through Slow PyrolysisProcess. In *IOP Conference Series: Earth and Environmental Science*, 1–7. doi:10.1088/1755-1315/ 913/1/012018.

Rahimpour, M. R., Aboosadi, Z. A., & Jahanmiri, A. H. (2012). Synthesis gas production in a novel hydrogen and oxygen perm-selective membranes tri-reformer for methanol production. *Journal of Natural Gas Science and Engineering*, 9, page 149-159.

Rahmani, A., Torkshavand, Z., Yari, K., Shabanloo, A., Foroghi, M., Gholazrkhogaste, B., and Najafi-Vosough, R. (2024). Preparation of Low-Cost Acidic Modified-Adsorbent from Agricultural Waste Palm Leaf to Remove Methylene Blue from Aqueous Solutions. *Current Research in Green and Sustainable Chemistry*, 9 (August). doi:10. 1016/j.crgsc.2024.100426.

Raksodewanto, A. A., Sudarmanta, B., Setiyawan, A., Priyanto, U., Fariza, O., Yarsono, S., Zufri, M. F., and Puspitasari, S. (2024). Pyrolysis of Oil Palm Trunk Biomass Using a Fixed Bed Reactor to Produce Raw Material for Bio-Carbon Black. In *IOP Conference Series: Earth and Environmental Science*, page 1–8. doi:10.1088/1755-1315/1344/1/012006.

Rasanen, E. L. S., Robertsen, L., Vuorenpalo, V. M., Karppi, A., Parviainen, K., Dahl, O., and Vantahalo, K. (2010). *Manufacturing of Microcelluse*. US9587038B2.

Rebenfeld, L. (1991). Hierachial Structure and Physical Properties of Natural Cellulosic Fibers. *MRS Online Proceedings Library*. 255: 399-404.

Reis, M. O., Zanela, J., Olivato, J., Garcia, P, S., Yamashita, F., and Grossmann, M. V. E. (2014). Microcrystalline Cellulose As Reinforcement in Thermoplastic Starch/Poly (Butylene Adipate-CO-terephthalate) Films. *Original Paper*. 22: 545-552.

Rosman, S., Sakinah, M. A., Ruzitah, M. S., and Iqbal, M. S. (2014). Effect of Fiber Surface Treatment on Mechanical Properties of Polylactic Acid/Empty Fruit Bunch Fiber Biocomposites. *Advanced Materials Research*. 875-877: 171-176. 10.4028/www.scientific.net/AMR.875-877.171.

Sabarisman, I., Suyatma, N. E., Ahmad, U. (2015). Pengembangan Film Nanokomposit Berbasis Pektin Dan Nanopartikel Zno Serta Aplikasinya Sebagai Pelapis Buah Salak Pondoh Utuh. *Master Tesis*. Institut Pertanian Bogor.

Sainz, C. B., Bustillos, R. J. A., Wood. D. F., Williams. T. G., and McHugh. T. H. (2010). Composite Edible Films Based on Hydroxypropyl Methylcellulose Reinforced with Microcrystalline Cellulose Nanoparticles. *Journal of Algicultural and Food Chemistry*. 58(6): 3753-3760. https://doi.org/10.1021/jf9033128.

Samat, N., Ghazali, N. N., Ahmad, Z., and Ali, F. (2018). Non-Acidic Method for Isolation of Microcrystalline Cellulose From Oil Palm Empty Fruit Bunch Fiber. *International Journal of Engineering and Technology*. 7(4). https://doi.org/10.14419/ijet.v7i4.14.27670.

Saragih, A., Oktaviani, O., Oktoria, W. D., Ekawati, R., and Saputri, L. H. (2022). Inovasi biopolybag ramah lingkungan dari tandan kosong kelapa sawit diperkuat dengan bahan isian sekam padi. *Jurnal Agro Industri Perkebunan*, 2(2), page 65-76.

Sari, S. A., Nurhayati, N., and Sunaryanto, R. (2022). Pengaruh Penambahan Effective Microorganisms (EM-4) terhadap Kualitas Limbah Cair Tahu dengan Teknik Aerasi. *Metrik Serial Teknologi dan Sains (E).*, 3(1), 36-41.

Sawit. (2012). http: fahriarief.wordpess.com. Diakses 18 Juni 2012.

Schwartz, M. M. (1984). Composite Materials Handbook. *Mcgraw Hill Book Company*. New York.

Sekar, S. M., Nagarajan. R., Selvakumar, P., Oluwarotimi, I. S., Krishnan, K., Mohammad, F., and Ayrilmis, N. (2024). Isolation of Microcrystalline Cellulose From Wood and Fabrication of Polylactic Acid (PLA) Based Green Biocomposites. *Journal of Renewable Materials*. 12(8): 1455-1474. https://doi.org/10.32604/jrm.2024.052952.

Senawi, R., Alauddin, S. M., Ruzitah, M. S., and Shueb, M. I. (2014). Effect of Fiber Surface Treatment on Mechanical Properties of Polylactic Acid/Empty Fruit Bunch Fiber Biocomposites. *Advanced Materials Research*. 875-877: 171-176. 10.4028/www.scientific.net/AMR.875-877.171.

Sidek, F. N., Saleh, S., and Saman, N. A. F. A. (2022). Kinetic Parameter Estimation for Pyrolysis of Empty Fruit Bunch Using Model-Fitting and Model-Free Methods. *Materials Today: Proceedings*, 57: 1241–47. doi:10.1016/j.matpr.2021.11.141.

Siyamak, S., Ibrahim, N, A., Abdolmohammadi, S., Yunus, W. M. Z. W., and Rahman, M. Z. A. (2012). Effect of fiber esterification on fundamental properties of oil palm empty fruit bunch fiber/poly(butylene adipate-co-terephthalate) biocomposites. *International Journal of Molecular Sciences*. 13: 1327-1346. 10.3390/ijms13021327.

Soon, V.S.Y., B.L.F. Chin, and A.C.R. Lim. (2016). Kinetic Study on Pyrolysis of Oil Palm Frond. *In IOP Conference Series: Materials Science and Engineering*, page 1–8. doi:10.1088/1757-899X/121/1/012004.

Srivasta, D., Ahopelto, J., and Karttunen, A. J., (2022). Thermodynamic Properties of Crystalline Cellulose Allomorphs Studied with Dispersion-Corrected Density Functional Methods. *Molecules*. 27(19): 1-11. https://doi.org/10.3390/molecules 27196240.

Sulhatun. (2005). Pemanfaatan Tandan Kosong Kelapa Sawit Sebagai Sumber Lignin. *Tesis Magister*, Sekolah Pascasarjana. USU. Medan.

Sun, R., Fang, J. M., Mott, L., and Bolton, J. (1999). Extraction and Characterization of Hemicellulose From Oil Palm Trunk and Empty Fruit Bunch Fibres. *Journal of Wood Chemistry and Technology*. 19 (1): 167-185.https://doi.org/10.1080/ 02773819909349606.

Sunardi., Wahyudi, R., Ula, S., Haryadi., and Alhamdi, A. A. (2025). Evaluation of Acoustic and Physical Properties of Composite Boards From Sugarcane Bagasse and Oil Palm Waste. *Polimesin*. 23(2).

Supian, M. A. F., Amin, K. N. M., Jamari, S. S., and Mohamad, S. (2018). Investigation on Pre-Treatment Process In Microcrystalline Cellulose (MCC) From Oil Palm Empty Fruit Bunch (EFB). *Chemical Engineering and Industrial Biotechnology*. 3(1): 87-96. https://doi.org/10.15282/jceib.v3i1.3877.

Surya, E., Michael, Halimatuddahliana, and Maulida. (2014). Impact and Thermal Properties of Unsaturated Polyester (UPR) Composites Filled with Empty Fruit Bunch Palm Oil (EFBPO) and Cellulose. *Advanced Materials Research*. 896: 310-314. https://doi.org/10.4028/www.scientific.net/AMR.896.310.

Taib, M. N. A. M., Salleh, A., Jamaluddin, N. A. N., Rasheed, T., Hussin, M. H., Julkapli, N. M., ... and Saleh, T. A. (2025). A review on recent developments of oil palm solid waste upcycling into green bio-based materials for sustainable building and construction applications. *Results in Engineering*, 107049.

Tan, Y. H., Tasirin, S. M., Saptoro, A., and Nolasco-Hipolito, C. (2019). Drying Characteristics of Oil Palm Frond Particles in a Spouted Bed Dryer. *In IOP Conference Series: Materials Science and Engineering*, page 1–8. doi:10.1088/1757-899X/495/1/012089.

Tang, Z. E., Lim, S., Pang, Y. L., Shuit, S. H., and Ong, H. C. (2020). Utilisation of Biomass Wastes Based Activated Carbon Supported Heterogeneous Acid Catalyst for Biodiesel Production. *Renewable Energy*, 158 page 91–102. doi:10.1016/j.renene.2020.05.119.

Then, Y. Y., Ibrahim, N. A., Zainuddin, N., Ariffin, H., and Yunus, W. M. Z. W. Y. (2013). Oil Palm Mesocarp Fiber as New Lignocellulosic Material for Fabrication of Polymer/Fiber Biocomposites. *International Journal of Polymer Science*, page 1–7. doi:10.1155/2013/797452.

Tursi, A. (2019). A review on biomass: importance, chemistry, classification, and conversion. *Biofuel Research Journal*, 6(2), page 962-979.

Uchegbulam, I., Momoh, E. O., and Agan, S. A. (2022). Potentials of palm kernel shell derivatives: a critical review on waste recovery for environmental sustainability. *Cleaner Materials*, 6(July). https://doi.org/10.1016/j.clema.2022.100154.

Umar, H. A., Sulaiman, S. A., Ahmad, R. K., and Tamili, S. N. (2020). Characterisation of Oil Palm Trunk and Frond as Fuel for Biomass Thermochemical. *In IOP Conference Series: Materials Science and Engineering*, page 1–10. doi:10.1088/1757-899X/863/1/012011.

Vedantu. (2025). *Cellulose: The Structure, Formula and Uses of Essential Plant Polymer.*

Wahyono, S., Sahwadan, F. L., and Suryanto, F. (2008). Tinjauan terhadap perkembangan penelitian pengolahan limbah padat pabrik kelapa sawit. *Jurnal Teknik Lingkungan, Khusus*, page 64–74.

Waigaonkar, S., Babu, B. J. C., and Rajput, A., (2011). Curing Studies of Unsaturated Polyester Resin in FRP Products. *Indian Journal of Engineering & Material Sciences*, Vol 18. Page 31-39.

Wang, T. Zhai, Y., Zhu, Y., Li, C., and Zeng, G. (2018). A Review of The Hydrothermal Carbonization of Biomass Waste for Hydrochar Formation: Process Conditions, Fundamentals, and Physicochemical Properties. *Renewable and Sustainable Energy Reviews*, 90, page 223–247. Available at: https://doi.org/10.1016/j.rser.2018.03.071.

Wirman, S. P., Fitri, Y., & Apriza, W. (2016). Karakterisasi komposit serat sabut kelapa sawit dengan perekat PVAc sebagai absorber. *Journal Online of Physics*, 1(2), page 10–15.

Wong, W. Y., Lim, S., Pang, Y. L., and Shuit, S. H. (2020). Science of the Total Environment Synthesis of Renewable Heterogeneous Acid Catalyst from Oil Palm Empty Fruit Bunch for Glycerol-Free Biodiesel Production. *Science of the Total Environment*, 727: 138534. doi:10.1016/j.scitotenv.2020.138534.

Yaakob, Z., Sukarman, I. S., Narayanan, B., Rozaimah, S., and Abdullah, S. (2012). Bioresource Technology Utilization of Palm Empty Fruit Bunch for the Production of Biodiesel from Jatropha Curcas Oil. *Bioresource Technology*, 104 page 695–700. doi:10.1016/j.biortech.2011.10.058.

Yang, P. M., Lin, Y. C., Lin, K. C., Jhang, S. R., Chen, S. C., Wang, C. C., Lin, Y. C. (2015). Comparison of carbonyl compounds emissions from a diesel engine generator fueled with blends of n-butanol, biodiesel and diesel. *Energy*, 90 page 1–8.

Yasin, M. H. M., Mamat, R., Najafi, G., Ali, O. M., Yusop, A. F. and Ali, M. H. (2017). Potentials of palm oil as new feedstock oil for a global alternative fuel: A review. *Renewable and Sustainable Energy Reviews*, Volume 79, Pages 1034-1049.

Yoricya, G., Aisyah, S., Dalimunthe, P., Manurung, R., and Bangun, N. (2016). Kelapa Sawit Dalam Sistem Cairan Ionik. *Jurnal Teknik Kimia*, 5(1) page 1–7.

Yoshi, L. A., Handayani, S., Ismojo, Enjarlis. (2025). Review: Potential of Oil Palm Empty Fruit Bunches Waste to Become an Ultrafiltration Membrane for Clean Water Treatment. *Journal of Bioresources and Environmental Sciences*, 4, (2), 88-101.

Younis Al-Ani, Y. M., Ridha, A. M., Abbas, T. K., Ahmadlouydarab, M., and Kamar, F. H. (2026). Modeling and Enhanced Biodiesel Production Using a Sustainable Green Catalyst Based on Palm Frond-Activated Carbon Supported with Zirconium Oxide. *Fuel*, 403 (February 2025). doi:10.1016/j.fuel.2025.136118.

Zainal, Mukhlis, M., Utomo, S., and Wijaya, M. (2025). Towards an Environmentally Friendly Palm Oil Industry: A Critical Review of Waste Reduction Policies by Indonesian Government. *International Journal of Sustainable Development & Planning,* 20(7).

Zaini, M. S. Md, Arshad, M., and Syed-Hassan, S. S. A. (2023). Adsorption Isotherm and Kinetic Study of Methane on Palm Kernel Shell-Derived Activated Carbon. *Journal of Bioresources and Bioproducts*, 8(1) page 66–77. doi:10.1016/j.jobab.2022.11.002.

Zakaria, M. R., Farid, M. A. A., Hafid, H. S., Andou, Y., and Hassan, M. A. (2024). Practical role of oil palm fronds in Malaysia's sustainable palm oil industry. *Industrial Crops and Products*, 222 (April). https://doi.org/10.1016/j.indcrop.2024.119753.

Zuhri, R. R. S., Manurung, R., Tambun, R., and Harahap, H. (2024). Application of Sodium Silicate from Palm Leaf Ash as Heterogeneous Catalyst in Transesterification of Palm Oil. *E3S Web of Conferences* 519, 02008 https://doi.org/10.1051/e3sconf/202451902008.

Index